P9-EMM-168

FORMERLY KNOWN AS FOOD

FORMERLY KNOWN AS FOOD

How the Industrial Food System Is Changing Our Minds, Bodies, and Culture

Kristin Lawless

ST. MARTIN'S PRESS
NEW YORK

www.stmartins.com

Design by Ellen Cipriano

Library of Congress Cataloging-in-Publication Data

Names: Lawless, Kristin, author.
Title: Formerly known as food : how the industrial food system is changing our minds, bodies, and culture / Kristin Lawless.
Description: First Edition. | New York : St. Martin's Press, [2018] | Includes bibliographical references and index.
Identifiers: LCCN 2017060758 | ISBN 9781250078315 (hardcover) | ISBN 9781466890565 (ebook)
Subjects: LCSH: Nutrition—Social aspects. | Diet—Social aspects. | Food industry and trade—Social aspects.
Classification: LCC RA784 .L39 2018 | DDC 363.8—dc23
LC record available at https://lccn.loc.gov/2017060758

First Edition: June 2018

10 9 8 7 6 5 4 3 2 1

To the people, who have the right to know.

"The public must decide whether it wishes to continue on the present road, and it can do so only when in full possession of the facts. In the words of Jean Rostand, 'The obligation to endure gives us the right to know.' "

RACHEL CARSON, *THE SILENT SPRING*, 1962

CONTENTS

INTRODUCTION

As a nutrition educator working with cardiologists in Manhattan, I've made a lot of house calls and seen a lot of kitchens. Usually it's not a pretty picture. One visit was to a woman I'll call Jennifer, who was struggling with her health and hoping to improve her diet. Coming into her kitchen, it was obvious what was wrong. Her shelves were full of highly processed food: Pop-Tarts, Frosted Flakes, and Oreos. She already knew she shouldn't be eating those foods, but she was shocked at my reaction when we opened her refrigerator and she showed off the contents: skim milk, fruit juice, a plastic bag of prechopped, prewashed broccoli florets, and Egg Beaters. These are all misguided ideas of how to eat well, I told her. The only consolation I could offer was that this is what most American kitchens look like. Jennifer needed to entirely rethink what she eats—as do most other Americans.

That night I went to dinner at a friend's house, someone I considered more enlightened about healthy eating, and I was a bit startled to see her kitchen stocked with Kashi cereal, soy milk, low-fat flavored yogurt cups, organic fruit and vegetable squeeze packs, and Annie's Macaroni & Cheese for her children. While these products are marketed as healthier options than Oreos and Pop-Tarts, the truth is they aren't much better than what I had seen on Jennifer's shelves that morning.

It's striking how quickly Americans have come to accept that it's normal for almost everything we eat to come out of a plastic bag, carton, or cardboard box. Indeed, since the mid-twentieth century, when the food and agricultural industries took over our food supply and began influencing dietary guidelines and so-called commonsense knowledge about food, we've completely lost touch with what actual food is. Even the whole foods we buy have deteriorated in nutritional quality and are contaminated with environmental chemicals. And our health has suffered in direct relation to these changes. Industry-influenced dietary guidelines have fundamentally altered even the way that many experts approach food and health. While I was giving a nutrition lecture at a prestigious cardiologist's practice in New York City a few years ago, I showed a slide defining the dangers of trans fats, those industrially produced fats in products like I Can't Believe It's Not Butter! or Benecol. I was explaining to the assembled group of patients and health practitioners that although we've long been told to avoid saturated fats like butter and replace them with these supposedly heart-healthy spreads, researchers now know that trans fats actually *cause* heart disease. I was in midsentence when a hand went up and an older man in the audience almost cried out in exasperation, "But that's the stuff my doctor has been telling me to eat for years!"

As it happened, this man's doctor was also in the room. I gently explained that this was relatively new information (this was 2011, and industrial trans fats weren't banned from the food supply until 2015), but, based on all the new evidence, people should avoid products with trans fats. The man's doctor was gracious and explained that nutrition was not part of his education or training, and he too was not aware of this cutting-edge nutrition information—which was why his office had brought me in.

I have spent more than a decade researching food, health, and nutrition, not only as a nutrition educator but also as an investigative journalist working strictly on food. Every year I interview dozens of biologists, chemists, toxicologists, and microbiologists doing groundbreaking research on the biological effects of the foods and chemicals that go into our bodies every day. Because of what I have learned from my research, I am increasingly alarmed when I go into people's homes and see the foods they are feeding their families. I've come to realize that we can't blame the eaters or even most of the doctors who mean well and are repeating official nutrition information from public health agencies and government dietary guidelines. But it is time to realize what our radically altered eating habits have done to our health. In just one hundred years a diet of packaged, processed, unhealthy food has become the norm for most of us. And even those of us who think we eat well are dining on fresh fruits and vegetables grown in depleted soils, often sprayed with pesticides, packaged in plastic that contaminates them, transported across great distances, and stored for weeks or months, all the while losing nutrients (like those broccoli florets). Because we have become reliant on these industrialized foods, our bodies are literally changing from the inside out, from the development of the fetus in the womb to the drastic changes in our microbiota—the trillions of bacterial cells that reside in and on us and are essential for good health. This dramatically new diet has also changed our brains, and has even altered our genes to the extent that what we eat now could affect our future children and their children, regardless of what they eat.

In fact, the explosion of research on the microbiota, a mother's diet and her breast milk, and the study of environmental chemicals that lace our food supply has suddenly called into question much of what we think we know about human health, the body, and our environment. Fundamentally, what we are learning is that over millions of years of evolution, humans have developed while in constant communication with our environment and the foods we eat. Until quite recently this system of response and adaptation served us well. Now, in less than one century, the industrialization of food production has completely changed nearly everything about the food we eat—and, by extension, our bodies.

In just a few generations we have managed to completely upend thousands of years of traditional knowledge about food, health, and nutrition. And no matter what your background, and no matter how healthy you think your food choices are, the saturation of our environment with industrial chemicals, along with the deterioration in the quality of even our most basic foods, represents the biggest public health crisis facing us today. We are all test subjects in a massive experiment, the outcome of which we are not likely to know for generations to come.

Both Jennifer and my friend—along with the rest of us—have to live in this radically new food environment. This book describes how we got here, what we are learning about its dangers, and how each of us might think about trying to live in it more safely.

PART ONE

How Did We End Up Here?

To many of us it seems like the food culture we have now is just the way it has always been—food ingredients concocted in labs, manufactured in faraway processing facilities, shipped around the country, loaded with preservatives and additives, and packaged in plastic wrappers or cardboard boxes. But this actually represents a deep rupture in the way we have eaten for the vast majority of our history. The American diet once consisted of actual food: meat from animals raised on a real farm, vegetables grown in nutrient-rich soil and plucked from a backyard garden, locally milled grains grown by a nearby farmer and baked into a fresh loaf of bread by a local baker or family member. Indeed, only in the last seventy-five to one hundred years have Americans become reliant on a highly processed and industrialized food supply.

But how did we get here? How is it that 70 percent of the population is now obese or overweight? How is it that children in staggering numbers are developing diseases once associated with old age, like type 2 diabetes and heart disease? The only real explanation for our soaring obesity rates and our decline in health is the drastic change to our diet during the past century, with the most dramatic changes occurring since the 1980s. The changes to our diets are quite profound and run deeper than you might imagine—from the way our foods are raised, grown, and produced, to the nutrients missing from them, to the addition of environmental chemicals and other additives. This represents nothing less than a seismic shift in our diets—much of what we eat now is not what it used to be.

Our Industrial Food Landscape

The Whole Egg Theory and How We Got Here

It's really quite simple. Any food you eat—anything that can be considered food—should be whole and intact, like the whole egg. For decades so-called experts have been confusing the public with misinformation about what makes up a healthy diet, when in fact, it's really not complicated. There is only one rule to remember: Any food that is whole should be part of your diet, and any food that is not should not be. Unfortunately implementing this rule has become infinitely complex in our industrial food landscape.

Do you even know what a whole food is anymore? Low-fat milk, nonfat yogurt, sliced white bread—these are not whole foods, but they've become everyday staples in most kitchens across America and beyond. Indeed nearly all packaged foods on grocery store shelves no longer qualify as whole foods. How did this happen?

The shift away from natural whole foods to packaged products really took hold in the American home in the 1950s and

1960s, when the government and the food industry started working hand in hand to develop the official dietary guidelines. Since that time, the public has been variously told to avoid eggs and eat only egg whites, use margarine and vegetable oils instead of butter or lard, drink low-fat or fat-free milk, use fats sparingly, eat lean protein, reduce salt consumption, make grains the basis of our diets, eat whole grains in the form of fortified commercial cereals and breads . . . and so on and so on. All these recommendations have one thing in common: they push you away from natural whole foods and toward the industry's processed foods. Just look at any grocery store shelf and you'll find corresponding health claims that do not reflect the wholesomeness of foods but rather the current dietary recommendations, which are influenced by the very producers of those foods—the ones touted as low fat, low cholesterol, without saturated fat, salt free or low sodium, and whole grain. These recommendations in turn have spurred entirely new markets for food manufacturers. It's key to keep in mind that manufacturers stand to make far more money from these processed foods, which use cheap raw ingredients like corn, soy, wheat, corn oil, soy oil, canola oil, sugar, and myriad additives and preservatives, than from simple whole foods. Your average box of cereal or package of sliced bread is a combination of these ingredients, which cost manufacturers little to produce. These products also have the advantage of lasting a long time on grocery store shelves and withstanding long transit times from factory to our kitchens. In contrast, whole foods such as freshly milled whole wheat, eggs, or butter are more expensive to produce and will spoil—which is what actual food does.

The very people we are supposed to trust to dispense nutrition information have endorsed the dietary guidelines that we see mirrored in the ingredients and marketing of packaged foods. The industry has wholly educated most registered dieticians employed in hospitals and doctors' offices and quoted as experts in the media. The world's largest organization of food and nutrition professionals, the Academy of Nutrition and Dietetics (AND), is the umbrella group that helps design the curricula for the schools that credential dieticians. But a 2013 report by Michele Simon, a public health lawyer, revealed "the food industry's deep infiltration" of AND. The organization's corporate sponsors are a veritable who's who of Big Food: Coca-Cola, the Hershey Center for Health and Nutrition, the National Dairy Council, General Mills, Kellogg's, PepsiCo, and Unilever. What's more, McDonald's, Coca-Cola, and other big-name food industry leaders spend a lot of money for booths and valuable floor space at the AND's annual expo, where they can promote the supposed health benefits of their products, which adds an air of legitimacy to their claims. After all, these corporations understand that dieticians are educating the public, so getting the dieticians' organization to endorse industrial food products is a major boon to business.[1]

Another organization that helps to shape the public's perception of what constitutes a healthy food is the American Heart Association (AHA). This group even puts what amounts to a seal of approval on products that meet its criteria for health. You've probably seen the red Heart-Check mark that says a product is low in saturated fat, cholesterol, and sodium. The AHA appears to be an independent group of doctors and researchers offering public health information, but deep conflicts of interest lie just below its surface. The organization boasts dozens of corporate sponsors in the food industry, including the Monsanto Fund, Walmart, Walgreens, Subway, and the Grocery Manufacturers Association (which represents big players in the food industry like Coca-Cola, General Mills, Hershey, Kellogg's, Kraft, and McDonald's).[2] The AHA has been endorsing low-fat, low-salt, high-fiber processed foods for decades and uses its Heart-Check program for food packaging to validate the healthfulness of packaged food products, many of which are of dubious quality. These products include Campbell's soups, Kashi cereals, Quaker Oats products, Butterball and Boar's Head meat products, and liquid egg white products, like Egg Beaters. Although the group said that it does not endorse any food, research that the AHA boasts about on its website shows that the Heart-Check mark is one of the most recognizable and trusted symbols on food packaging. Eighty-three percent of consumers are aware of the symbol and 63 percent trust the Heart-Check mark most.[3] That symbol comes at a price for food manufacturers, who pay the AHA an annual fee to use it. Payments are tiered according to the number of products and the companies' sales revenue and range from $3,000 to $6,000 per license per product annually. As I write this, 960 products bearing the Heart-Check mark are on the market.[4]

Trusted recommendations from groups like the Academy of Nutrition and Dietetics and the American Heart Association have come to replace common sense about food and nutrition; most people still assume that eating low-fat and low-salt food products is good for the heart and overall health. Truthfully neither recommendation is based on sound science, and both have come at a tremendous cost to the public health. Indeed one of the pitfalls of nutrition science of the past half century or so has been the vilification of entire food groups like fat, protein, and carbohydrates—rather than an examination of the vast complications of our current food landscape with its many problematic ingredients and its ultraprocessed nature.

One of the most prominent examples is our misunderstanding of fats, vilified for decades, especially by the AHA. This has had dire consequences for our health, affecting all of us. The demonization of saturated fat, such as butter, goes hand in hand with the addition of artificial trans fats to our food supply. For decades the AHA tacitly endorsed the consumption of trans fats such as margarine in place of saturated fats as a way to reduce the risk of heart disease—however, we now know that trans fats damage your arteries and actually *cause heart disease.* One Harvard nutrition expert has said that trans fats are the biggest disaster in food-processing history, resulting in thousands of deaths every year. The American Heart Association now has backtracked on trans fats but still recommends reducing your saturated fat intake in general and using polyunsaturated vegetable oils instead. But that advice is based on outdated science. The four studies that the AHA uses to support this claim all date to the 1960s and 1970s, which makes them antiquated by nutrition science standards. In 2017 a major study in *The Lancet* that looked at more than 135,000 people across eighteen countries and five continents found no association of fats of any kind, total fat intake, and risk of cardiovascular disease or cardiovascular disease mortality. In fact, the study found that a higher saturated fat intake is associated with a lower risk of stroke.[5] When I asked the AHA about this study, its staff referred me to those outdated studies that back its original claim. For decades the group also advocated eating carbohydrates, such as pasta, breads, and other grain-based snack foods, instead of fats, a recommendation that is at least partially responsible for our surging obesity and diabetes rates—although now the AHA admits that the science no longer supports that advice. The Heart-Check mark nonetheless continues to grace the packaging of hundreds of processed low-fat products on grocery store shelves. One of the most egregious examples is Egg Beaters, an egg white–only product used as a replacement for fresh, whole eggs. The notion that this is a healthy alternative is inane—which brings us back to the whole egg—or my Whole Egg Theory.

An egg is a perfect food. It is a rich source of nutrition that is balanced and complete. After all, it holds all the nutrition needed to develop a chicken. Humans have been eating eggs for at least four thousand years—since we domesticated the chicken. (In contrast, we've been eating Egg Beaters for about forty-five years. As a rule, anything that we've been eating for less than one hundred years should be considered experimental.) The demonization of eggs in general, and the concept of eating only egg whites, is a clear illustration of the way we have undermined the basic goodness of the natural whole foods that humans have been eating for thousands of years. It makes absolutely no sense to discard egg yolks—they are the nutritional powerhouse of the egg, rich in B vitamins, vitamin K, selenium, vitamin D, and protein. In fact, the ratio of amino acids in a whole egg is as close to ideal for human nutrition as any food can be.[6] Egg yolks are high in nutrients that promote heart health, such as betaine, which reduces homocysteine. Too much homocysteine in the body damages blood vessel walls and is linked to increased risk for heart disease, osteoporosis, Alzheimer's disease, and cancer—as well as neural tube defects.

Eggs are also rich in choline, which is crucial to every cell membrane in the body and for brain health and function. Choline is especially important to the development of the fetal brain, which makes eggs a vital food during pregnancy and breast-feeding. Eggs are high in antioxidants such as glutathione, which helps fight cancer, and the yolks are rich in carotenes, lutein, and zeaxanthin, important for eye health and for preventing colon cancer. Eggs have one of the highest levels of biotin—a B vitamin crucial for healthy skin, hair, and nerves—of any food. Biotin also helps with the digestion of fat and protein.[7] The whites contain valuable nutrition too—but they work in concert with the egg yolk in a perfect synergy that you might say was designed that way by nature.

The abominable egg white omelet comes from the idea that because the egg yolk contains cholesterol, eating only the egg white is healthier—all the protein, none of the fat. Behind this is another major misconception about our understanding of foods and their effects on our bodies. The concept was that if you eat foods high in cholesterol, that cholesterol will then clog up your arteries. There's just one problem: dietary cholesterol, that is, the foods you eat that contain cholesterol, is correlated only weakly with the amount of cholesterol in your blood. Indeed, your liver produces cholesterol, because it is vital to our bodies, whether we consume it in foods or not. This has finally become accepted in mainstream medicine and nutrition research, so in 2015 the United States Department of Agriculture dropped its recommendation to limit cholesterol consumption from the official dietary guidelines.[8] Most people did not get the memo, however. Whenever I talk about the health value of eggs, people often say to me, "But what about the cholesterol?" Sadly, undoing decades of demonizing whole foods like eggs is probably going to take an equal amount of time—so the egg-white omelet persists.

We like to think we can outsmart nature by separating food into its constituent parts to make it better in some way—but in fact we undermine our own interests in doing so. My Whole Egg Theory applies to all other foods and can serve as a guiding principle for figuring out what—and what not—to eat. The idea of improving natural whole foods by removing, separating, manipulating, and adding to them has corrupted our entire food supply in ways that have completely changed the definition of *food*.

This is often comical because once these constituents are removed from foods, they are added to other foods and sold to us under the guise of better health. Some of the worst examples are Pepsi with added fiber and fake butter spreads with omega-3 fatty acids. But you cannot eat an extracted or synthetic element of a whole food and expect to get the same health benefit as eating the whole food itself. Vitamins, minerals, fatty acids, and all nutrients exist within the synergistic matrix of the food, and those mutually beneficial nutritional relationships cannot be replicated in a lab.

This backward logic of the food industry has multiple angles. Another angle we often hear is that because chemicals already exist in nature, chemical additives are perfectly harmless. But the difference between an ingredient in its whole state and one that is extracted from a plant, isolated in a lab, and then added to our foods is vast. Take the example of the guar plant, something people in India have eaten for thousands of years, and guar gum, which is added as an emulsifier to many organic and natural food products, like peanut butter or ice cream, to keep them from separating. Crucially these so-called natural emulsifiers act differently in the body than the whole plant does, in part because the concentrations are vastly different. In fact studies using animal models have shown that guar gum has strong effects on the bacteria in our gut (the microbiota), damages the lining of our intestines, and leads to a variety of gastrointestinal issues.[9] The claim that an ingredient is safe because it has been extracted from a natural source has been used to defend everything from aspartame to trans fats because components of these industrial ingredients are also present in whole foods. But this is industry spin—simply because something is found in nature doesn't mean it belongs in extracted form in our food.

We need more nuance and specificity to truly understand the scope of the problems we have created, especially in our language—when we refer to *fats*, what do we mean? Are we talking about the kinds of natural fats found in butter, dairy, or eggs? Or are we talking about the processed liquid vegetable oils found in packaged foods? Our bodies respond differently to, on the one hand, a fat found naturally in a whole food and, on the other, a liquid extracted from a bean or seed that becomes part of a cracker or cookie. In other words, the soybean oil in a cracker is not comparable to the fats in butter, milk, or cheese.

What's more, whole foods contain more than one type of fat—they are a combination of various types of fats and fatty acids. The fats in foods are often broken down into four categories: saturated fats, trans fats, monounsaturated fats, and polyunsaturated fats. Lard, perhaps the most demonized fat of all, is assumed to be all saturated, but it is actually 40 percent monounsaturated fat, 11 percent polyunsaturated fat, and 39 percent saturated. This

FATS

The four main types of fats found in food are saturated fat, trans fat, polyunsaturated fat, and monounsaturated fat. Each category has dozens of fatty acids, and the way they act inside our bodies is complex and in some cases not fully understood—but the main types that I am discussing here are:

- **Saturated fats**—mainly found in animal foods like cheese, butter, milk, beef, and pork. They are also present in large amounts in coconut and palm oils.
- **Trans fats**—a type of unsaturated fat, both artificial and naturally occurring (I discuss them in greater detail in chapter 2). Artificial trans fats are mainly found in products like margarine and shortening, as well as in an array of processed foods like cookies, crackers, and other snack foods. The term *partially hydrogenated* on labels means trans fats are present. Many restaurants and bakeries also use them. Artificial trans fats are made by heating and processing liquid vegetable oils.
- **Polyunsaturated fats**—found in high amounts in sunflower, corn, soybean, and flax oils. They are also present in nuts and seeds like walnuts and flax seeds and in fish. Omega-3 fatty acids are a type of polyunsaturated fat.
- **Monounsaturated fats**—found in olive, peanut, and canola oils. They are also present in avocados, almonds, hazelnuts, and pecans, as well as pumpkin and sesame seeds.
- Keep in mind that most foods are a combination of various types of fats and fatty acids. Only rarely is a food all saturated fat or all unsaturated fat.
- Naturally occurring saturated fats are healthful, but animal fats vary in their composition based on what the animal ate. For example, because of the plant matter the animals eat, beef, dairy, and butter from grass-fed cattle tend to

have higher amounts of the beneficial omega-3 fatty acids than their industrially raised counterparts. Feedlot cattle don't eat much plant matter and instead eat mostly corn, soy, and other grains.

- After decades of use, we now know that artificial, industrially produced trans fats cause arterial damage and heart disease. Small amounts of trans fatty acids occur naturally in some foods, like dairy or meat from grass-fed cattle, but these are not harmful.
- Both polyunsaturated and monounsaturated fats are healthful when consumed in whole foods, but the abundance of them present in soybean, corn oil, canola oil and other vegetable oils that lace our food supply are problematic for health.

Fats are a prime example of why nuance is important when discussing food, health, and nutrition.

means it is actually full of what experts have been calling "heart-healthy fats," like the monounsaturated kind found in olive oil. And that much-maligned steak contains a balanced mix of fats as well: an average-sized steak contains 21 grams of saturated fat, 23 grams of monounsaturated fat, and 2.2 grams of polyunsaturated fats. It's true that some foods are predominately one type of fat: butter and coconut oil are nearly all saturated fat, and olive oil is mostly monounsaturated fat. But trying to simplify the complex nature of our foods in the name of creating easy-to-follow guidelines for the public has backfired and undermined our understanding of food and nutrition—not to mention our health.

Had we not industrialized our food supply, all this would be moot. If we still ate only whole foods, we wouldn't give even a passing thought to the types of fats or fatty acids in our foods—we'd just eat the eggs, the butter, or the avocado because those are the foods we eat. Therefore the very idea of nutrition and food science is essentially a by-product of the processed food industry—no wonder it seems impossible to get it right. How can we expect to understand the nature of our foods and how they act in our bodies when we are not really eating food anymore?

Our notion of food has been so corrupted that even when researchers are testing different diets in lab animals, they are often feeding them a highly processed, refined food diet as a baseline. So when they test a supposedly high-fat diet, do the results reflect the high fat content or the processed nature of the feed? And until the dangers of trans fats were widely accepted, scientists were lumping trans fats and saturated fats into the same category when trying to determine the effects of a high-fat diet. Such studies have led to gross oversimplifications about components of our diets, as well as outright fallacies. If a so-called dietary expert tells you to eat low-fat foods, or a low-fat diet, you can be sure that person has no idea what they are talking about. The other possibility, of course, is that they are mouthpieces for an industry (either wittingly or unwittingly) that profits from confusing and contradictory nutrition advice as well as the thousands of products that fit the low-fat description. Where is the profit margin in saying that a whole food like butter is healthy? It is far more profitable to market the hundreds of butter-replacement spreads and thousands of low-fat foods on grocery store shelves. Suddenly we've gone from one simple food to thousands and thousands of marketable foods under the guise of better health.

In the spirit of nuance and specificity, here's a comprehensive definition of *processed foods* to keep in mind: products mechanically separated, altered from their original state, and then combined with various additives. They are often based on highly refined grains like wheat and corn that have been stripped of their more nutritious components and then laced with additives like artificial flavors and preservatives, fats in the form of liquid vegetable oils, sugars both artificial and natural, and sodium. Many of these products are then fortified with synthetic vitamins. They are also subject to contamination from chemicals in food packaging or during the manufacturing process, that are known to interfere with our hormonal, metabolic, and neurodevelopmental systems. Finally, if made with conventionally grown crops, these products contain residues of pesticides, herbicides, fungicides, and insecticides and, in the case of foods derived from animals, growth hormones and antibiotic residue. Most foods on grocery store shelves fit this profile: crackers, chips, cookies, cereals, sliced bread, yogurt, juices, macaroni and cheese, frozen dinners, and granola bars, to name a few.

Keep this definition of processed foods in mind as I demonstrate that the word *food* no longer applies to much of what we put into our bodies, which means not only has our food changed drastically but our bodies have too.

Indeed, industrial food has made us—and future generations—different human beings, all within the past seventy-five to one hundred years. From the changed composition of the soil in which we grow our foods to the additives in our kids' snack packs to the plastics that touch nearly all our foods, thousands upon thousands of substances make their way into our bodies that never did before.

We've all heard the news ad nauseam that Americans are growing heavier than ever, but it's not just the size of our bodies that is changing; what we eat is actually changing our internal makeup. The food industry is shaping the taste preferences of babies, children, and even the unborn, creating desires for unhealthy foods. What a mother eats while she is pregnant and breast-feeding appears to have essentially permanent long-term health effects for her baby. In addition, babies fed formula—their first processed food—suffer long-term health consequences. And both formula and breastfed babies now have a compromised gut microbiota due in part to the extinction of a crucial protective bacterial strain that is missing from the industrialized world.

This bacterial extinction is also seen in the adult microbiota, with important strains now missing from our modern world, thanks to our diet of highly processed food products. And because our gut bacteria also seem to influence brain health and function, these are likely compromised too. The changes to the gut appear to be passed on to future generations as well—you inherit your microbiota from your mother, and if her gut was imbalanced, then yours probably is too. And the addition of thousands of environmental chemicals have so saturated our food supply and our environment that we cannot escape them—no matter how well we eat.

It is not only what is added to our foods but what is missing as well. Even our most basic whole foods have changed dramatically as a result of the industrialization of our food supply, often resulting in deficiencies. We are even seeing widespread malnourishment in the population of largely overweight and obese people—individuals who are overfed but undernourished.

Because we've been eating this way for only about 75 years of our 200,000-year existence on this planet—or 0.000375 percent of human history—the only way to describe the industrial food system and its effects on our health is as a short-lived experiment that is only beginning to yield some results. Not only is our industrial food system an experiment, but the entire history of farming is relatively new and untested for long-term success. The dawn of agriculture marks what some historians regard as the beginning of our eventual downfall as a species because the population explosion that accompanied our agrarian lifestyle is the basis of climate change. Still, the newness of agriculture itself is nothing compared to the serious threat to our health, our bodies, and our way of life that the industrial food system represents. And nothing is a greater threat to our survival as a species than the way we produce food and the way we now eat.

The statistics that describe the sorry state of health among Americans (and, more broadly, the world) provide a glimpse into some of the frightening results that this industrial food experiment is yielding. More than one-third of all Americans are obese; add to them those who are overweight, and more than 70 percent of all Americans have a problem with excess weight. Globally, more than two billion people are overweight and more than six hundred million are obese. Worldwide, obesity has doubled since 1980; today more of the world's population lives in countries where being overweight and obese kills more people than being underweight. And for the first time in our history, chronic disease kills more people than infectious disease around the world. In the United States the rates of chronic disease are unprecedented. Fully half of the entire population has either diabetes or prediabetes, which is a cluster of conditions that includes high blood pressure, abdominal obesity, high triglycerides, low amounts of high-density lipoproteins (HDL, the "good cholesterol"), and high fasting blood sugar. The medical community once referred to type 2 diabetes as "adult-onset" but has since dropped this descriptor as rates of children with the disease have skyrocketed in recent years. From 2001 to 2009 the disease among Americans aged ten to nineteen shot up by 30 percent. More alarming still, doctors are seeing a stark increase in colon and rectal cancer among twenty- and thirty-year-olds. People born in 1990 have double the risk of colon cancer and quadruple the risk of rectal cancer than someone born in 1950.

So, despite the growing awareness of the need for "healthy eating" and what's been termed the obesity epidemic, the state of our foods and our health has only worsened.

Who's to blame? The food industry claims it is all about "personal responsibility" and makes inane statements like "all foods in moderation," or it tells people to "move more" and "eat less." Stigmatizing people who are overweight or obese is one of the last prejudices considered acceptable by otherwise thoughtful people, who tend to see obesity as a moral failure or a lack of willpower. But the emerging science about metabolism and endocrinology says otherwise—all calories are not created equal, you can't exercise your way out of a bad diet, and the adulterated and contaminated nature of our foods is changing metabolic functioning regardless of caloric intake. Our rapidly deteriorating and contaminated food supply is to blame for these unprecedented surges in obesity and chronic disease. There is simply no other explanation.

But that's not the whole story—yes, our foods have changed, but so have our food habits and behaviors, touching nearly every facet of American life and culture: longer work hours, shifting gender roles in and outside the home, reductions in real wages, and the decline of the middle class; lack of child care, lack of paid parental leave, and changes in our geographic relationship to foods; and the gentrification of urban centers, where wealthier areas abound with healthy, whole food options at farmer's markets, gourmet shops, and well-stocked supermarkets, while poorer sections of our cities lack even one grocery store, leaving residents to rely on the cheap processed foods found at corner stores, drug stores, or liquor stores. Many of these factors are well beyond our individual control and in fact largely determine and limit our so-called food choices every day.

And it is crucial to point out that some Americans are affected more than others. African American women have higher rates of obesity than any other group in the United States. Roughly four out of five African American women are now overweight or obese, African Americans are 1.5 times as likely to be obese as whites, and African Americans are 30 percent more likely than whites to die young from heart disease. Seventy-eight percent of Mexican American women are overweight or obese; Latinos are 1.2 times more likely to be obese than whites, and Latinos are 50 percent more likely than whites to die of diabetes. One in three American children born in the year 2000 is expected to develop diabetes at some point, but one in every two African American and Latino children is expected to be diagnosed with diabetes.

Race-based health disparities are inextricably linked to our food supply. Because of neighborhood segregation, businesses like grocery stores left low-income neighborhoods generations ago, creating the much-discussed food deserts. Residents of these neighborhoods are also exposed to greater amounts of environmental chemicals because heavily processed foods and drinks have higher amounts of them than fresh, whole foods.

These external factors throw the idea of personal responsibility into a tailspin. How can we blame the individual when so much is now out of our control? And when we add the internal factors—that is, the effects that our processed food diets are having inside our bodies—the question of choice becomes even more complicated. And the idea of choice is largely illusory for yet another reason: the industry presents its choice in the form of thousands of products—almost all unhealthy—controlled by just a handful of corporations. So, while it might appear that we have endless choices, industry dictates the parameters of those choices. Furthermore, decisions about what substances end up in our foods—many of them untested—are made by the industries involved in the production of those substances.

At the same time, the food industry has been brilliant at framing the concept of whole foods as elitist or strange—the antithesis of true American fare. And with the current interest in healthy foods, whole foods, and artisanal foods, the industry has mounted an even more aggressive attack, pitting liberal elites and their unusual foods, such as kale or quinoa, against the average Joe, who just wants to eat his McDonald's burger, drink his supersized Coke, and be left alone by a "nanny state" bent on telling everyone what and how they should eat. The trouble is, many Americans are buying this outright manipulation and inversion of reality. The government actually plays a relatively small role in telling people what to eat, while Big Food pumps billions of dollars into advertising and marketing. And, even more alarming, the regulatory agencies of the government, the Food and Drug Administration and Environmental Protection Agency, are barely regulating much of anything in our foods.

Yet the "food movement," the opposition to the industrial food system that has taken shape in America, has no remedy for glaring disparities across racial, class, and cultural lines. The food movement's solutions often mirror the food industry's mantra of personal responsibility and consumer choice, which completely skirts the institutionalized and systemic lack of access to healthy whole foods for large segments of the population. And even those who follow the advice of the food movement and buy organic or shop at farmers markets, cannot escape the pervasive and largely unregulated toxic ingredients in the food supply, nor can consumer-based solutions address the ways our modern diet is changing our bodies, leading to the staggering rates of diet-related disease, suffering, and death. As such, many food movement leaders have fallen prey to the language and ideology of the industry that they seek to combat, often echoing the false notion that personal responsibility and choice are more consequential than these larger and more complicated factors.

An unintended consequence of the growing food movement and the explosion of foodie culture has been the widening divide between those who have the money and access to buy healthy alternative foods and all other Americans, who must rely on processed, unhealthy food. Gourmet markets or farmers markets that carry rarefied heirloom or heritage ingredients, serve as the symbolic alternative to the typical and ubiquitous American fare. But these alternatives are so far out of reach for most Americans that we essentially have one food system for the food elites and one for everyone else. Indeed, the rich and famous have usurped the language of whole foods, organics, and fair trade and turned it into a parody of itself. But in truth these terms represent real principles that all people, regardless of economic or social status, should be able to afford and embrace.

To truly challenge the industrial food system, we must go beyond personal and consumer choice. We can't simply continue to insist that individuals reject processed foods and cook at home. That message is incomplete and ineffective. We have to collectively reorganize the way our food system works. This will require a true democratization of the food supply, that is, the participation, to some degree, of a majority of Americans in the production and distribution of our food—as was the case not so long ago. In the 1850s more than half of the U.S. population lived on farms, and more than 60 percent of the total workforce was farm labor. Now the percentage of the population that works on farms hovers around one percent, and the industrial food system has hijacked this fundamentally American way of life. This doesn't mean most of us will be farmers again, but we must be engaged with our food supply in some direct way. This could look like local, small-scale farming or gardening, participating in a member-owned food co-op, getting involved with community food production and education, or direct action protests against Big Food and Big Ag. We must reframe the value of growing, producing, distributing, and preparing food so that we once again understand its true value—this will be critical to our survival.

A true democratization of the food supply will foster a collective rejection of the industrial food supply. To get there, we need to unravel the persuasive and tangled narrative about American food, freedom, personal responsibility, and choice that the food industry has been so great at spinning to its advantage. We will also need to look more seriously at the economic factors forcing so many of us to be dependent on cheap, convenient foods. If we prioritize and collectively value the time and skill it takes to produce food, from growing it to cooking it, the health of this nation will change in step with those priorities and we will see that the industrial food system, and the current alternatives, are part and parcel of an economic system that is unsustainable and unfair for all but those at the very top.

This book is meant to arm you with the information you need to keep you and your family healthy and safe in a food landscape that is riddled with landmines, the unseen and untold harms that too many of us don't even know exist. This book is also a call to action, a declaration of all that's gone wrong with our food system, what we're up against, and how we can collectively fight for real change. Our entire food landscape has changed in just a couple of generations, and we must ask what we have lost in the march toward a completely industrialized food supply. What may seem like progress to some has regressed our health profoundly. Indeed, many defenders of our modern food system argue that our longer life spans are clear evidence that we are faring better than ever before. But that too is changing: for the first time in two centuries the current generation of children is expected to have a shorter life span than their parents.[10] And in December 2016, the Centers for Disease Control and Prevention reported that, for the first time since 1993, life expectancy had declined in the United

States, and six of the ten leading causes of death are directly related to our diets: heart disease, cancer, stroke, Alzheimer's disease, diabetes, and kidney disease.[11] For those who may be living longer, we must ask: Are we living better? Living into our seventies or eighties with diabetes, heart disease, or cancer while heavily dependent on medications and treatments is far from ideal.

It is hard to overstate just how dramatically the convergence of new technologies and the powerful interests of big business is altering how we raise, grow, produce, eat, and relate to our food. If we don't take drastic and radical steps to completely change how food is produced and consumed in this country, we are looking at a frightening future. The current estimate is that one-third of all greenhouse gas emissions come from the industrial food system,[12] and it is clear that how we grow food in this country is just as inextricably tied to the climate crisis as it is to the health crisis. We cannot continue eating the way we do without destroying our planet and ultimately ourselves. The industrial food system has changed human beings, both in our culture and habits and in our physical and mental health.

The good news is that if our industrial food system is at the root of these environmental and health crises, many of the solutions will be found in reframing our agricultural systems and reevaluating the way we feed ourselves—which will mean a complete overhaul of today's industrial farming and food models. This book is a wake-up call to all Americans. The time has come to change course and demand action for our future.

2

What Are We Actually Eating?

Industrial Food and Our Health

My grandmother, who lived to be ninety-seven, was always skeptical of the benefits of organic foods. She thought it a marketing ploy to get people to spend more money. After all, she said, "I didn't eat organic, and look at me!" When she died in 2015, she'd had few major health issues, no chronic diseases, and lived on her own in downtown Philadelphia for more than twenty years. But here's the catch: for at least the first several decades of her life, she did eat organic. Of course, no one called it organic, but by default her food was just that. Her diet in the 1920s, '30s, and '40s was, by and large, chemical free, grown in healthy soils without synthetic fertilizers, and it was fresher and probably much more nutrient dense than the foods we eat today.

My grandmother never fell for the hyped notion that processed foods offer great advantages over the foods she and her family grew up with. She and her brother, my uncle Ben (who lived to be ninety-three), always ate whole foods and never missed a chance to teach my sister and me about the virtues of, say, broccoli, string beans, eggs, butter, or a roasted chicken. And this was the 1980s and '90s—a time when the processed food industry kicked its production and marketing into high gear.

But my grandmother was immune to the marketing, given her lifelong experiences eating whole foods. Imagine, if you will, that it's 1935: You wake up early in the morning and the milkman has just left a delivery of fresh milk in glass bottles for the week. You bring in the milk and make breakfast: scrambled eggs, a glass of milk, and fresh bread bought from the bakery in your town, which you eat with a nice slathering of butter. Even in the 1980s and '90s, my grandmother served us eggs cooked in butter along with Jewish rye (with some more butter) and fruit—none of the chemical-laden breakfast cereals, sugary breakfast pastries, and fake butter spreads that grace the dining tables of most Americans today. Sometimes we even had those eggs cooked in bacon fat, just like Uncle Ben would cook for my dad when he was a kid—a traditional breakfast going back generations. "Don't believe what they say," Ben would tell my dad. "Animal fats are the best thing in the world for you." Uncle Ben was on to something here—and in fact was passing on the wisdom of generations past. Scientists have an abundance of evidence that traditional people from cultures around the globe thrived on animal fats with no signs of heart disease or any other "diseases of civilization," such as diabetes or cancer. Indeed, recent research exonerates animal fats, or saturated fats, from causing harm to our health. An exhaustive 2014 analysis of more than eighty studies, including evidence from twenty-seven peer-reviewed studies (the gold standard in research) and involving more than half a million people found no evidence that eating saturated fats causes heart disease or other cardiac events. A 2016 analysis of statistics from forty-two European countries found that high consumption of total fat and animal protein was linked to lower risk of cardiovascular disease. And, as I mentioned in chapter 1, in 2017 *The Lancet* published a major study that found no correlation between fat of any kind and increased risk for cardiovascular disease or mortality—in fact saturated fat, like the kind in butter, was associated with a decreased risk of stroke.[1]

So what did you have for breakfast? If you're like most Americans, pressed for time and rushing to get to work, you did one of two things: you made coffee and ate a quick bowl of cereal with milk, or maybe you stopped for coffee and a breakfast sandwich or pastry on the way to the office. The vast differences between my grandmother's breakfast and yours, separated by about eighty years, may not be immediately obvious—but analyzing them makes clear that nearly everything we eat today is different from what Americans ate just two generations ago. This is not to romanticize some perfect past in our food history but rather to show just how drastically the content of our foods has changed, which, as you'll see, is undeniable.

I'll begin with that seemingly innocent bowl of cereal. When my grandmother received her delivery of milk in the morning, she was pretty much guaranteed that it was just milk. But the milk in your fridge today is substantially different from hers. Her milk came in glass bottles. The milk we drink today comes in either a plastic bottle or a lined paper container, and both raise questions about safety. The plastic bottle is most likely made with bisphenol A (BPA), a synthetic compound used to make plastics and epoxy resins that is alarming experts who have found that it acts like a hormone and interferes with our endocrine systems. BPA appears to be an "obesogen," which means that it causes weight gain no matter the amount of calories you consume. And the stuff is everywhere—the Centers for Disease Control and Prevention say that "nearly all Americans" harbor detectable levels of BPA in their bodies. And its obesogenic effects are not the only cause for concern: more than one thousand studies link the substance to problems with fertility, certain cancers like breast and prostate, cardiovascular problems, and impaired brain development in fetuses and babies exposed to BPA.[2] And many of these findings are for low-dose exposures—in some cases endocrine-disrupting chemicals like BPA cause greater disruption to our systems in low doses than in high doses.

Milk that comes in a paper carton may be only marginally better, because the carton is most likely coated with a polyethylene resin, a petroleum-derived plastic that also appears to interfere with our hormones. All plastics have what is known as "estrogenic activity," meaning they can mimic or interfere with the hormone estrogen, potentially altering our bodies in unknown ways.[3] This is no small concern because most modern processed foods are packaged in some form of plastic, and we are exposed to it every day in multiple forms.

What about the milk itself? Your milk is significantly different than milk was only forty or fifty years ago. Back then, milk processing involved little more than milking cows and heating the milk to kill any harmful bacteria (although many people on farms drank raw milk with no problem). The milk we drink today is a far cry from what comes out of a single cow. In fact, the milk in your fridge probably came from many different cows, because holding silos combine tens of thousands of gallons of milk from many different farms. The milk goes through a cream separator to create two products: cream and skim milk. Then various percentages of cream are added back to the skim milk to create whole and low-fat milk. The milk is then homogenized by passing it through small holes at high speeds to create a uniform texture and prevent the cream from separating and rising to the top. It's then pasteurized by being heated to at least 145 degrees. In some states milk wholesalers add nonfat milk solids and milk protein concentrates to thicken the milk and give it a better texture. Then the wholesalers add synthetic vitamins A and D.[4]

This process is problematic for many reasons. The various additives, and the processes involved in producing this basic staple, essentially guarantee that we're getting a mixture of substances from all over the country and perhaps the world. Without a single origin—say, one particular farm—any food-borne illness resulting from contamination becomes nearly impossible to trace. Little ability to trace means little to no accountability. Not only is the milk coming from various farms in the United States, but other substances added to the milk, like milk protein concentrate, often come from China, New Zealand, and Mexico, where food safety and regulation requirements are questionable. This means that, despite their widespread use in our food supply, milk protein concentrates are unregulated in the United States. In fact, the Food and Drug Administration (FDA) does not even classify them as generally recognized as safe (as I will show, this classification is no assurance that a substance is safe).

Milk protein concentrates are made by ultrafiltration—forcing milk through a membrane to remove some of its lactose; the concentrates are lower in carbohydrates and higher in protein than other milk solids, which makes them ideal additions not only to milk but to processed foods such as protein bars, drinks, and processed cheeses. Dairy processors and food manufacturers use milk protein concentrates because they are inexpensive substitutes for real milk. The FDA says it has little concern about the safety of milk protein concentrates because they are treated with heat during pasteurization and drying. But the way they are made raises some concern about what they do once they are in our bodies. They are similar to another ingredient that is often added to commercial milk: nonfat milk solids. Both ingredients are made through ultrafiltration of skim milk and spray drying, followed by the application of varying degrees of high heat to evaporate any remaining water content. Needless to say, these are not whole foods. Furthermore, the heavy amount of processing and the application of high heat probably damage any remaining fat molecules in the milk. Damage to fat and cholesterol molecules by heat results in oxidization, and oxidized cholesterol has been shown to cause atherosclerotic plaques, injuring our cell membranes and resulting in damage to the arterial linings, eventually leading to heart disease. (Your best bet, short of living on a farm and milking your own cow, is to find a local farmer who uses "lowest-temperature" pasteurization and sells "cream on top" whole milk, preferably in glass bottles.)

Fred Kummerow was a researcher who died at the age of 102 in 2017 and in the late 1950s was one of the first scientists to sound the alarm about the dangers of trans fats. According to Kummerow, oxidized fats are just as bad for our bodies as trans fats. He published his first paper about the harms of trans fats in 1957, about the time they became prevalent in our food supply (as partially hydrogenated vegetable oils like those found in margarine), and only since 2007 have the medical and scientific communities acknowledged that he was right all along. "Trans fats and oxidized fats are bad things," he told me in a 2015 interview. "Oxidized fats damage the wall of the artery and atherosclerosis is produced; people who get that kind of damage need a coronary bypass operation or they will die. This is the result of exposure to any oxidized fat—it can be egg powder, or from [fats used for] frying—a number of different oxidized fats will do the same thing. Any [fat] that has been dried for a while with heat is going to be oxidized."[5] Dried fats like milk protein concentrates, nonfat milk solids, and egg powder lace our food supply. And most restaurants are using poor quality liquid fats for frying.

High heat and processing applied to delicate fats make the safety of homogenization and pasteurization questionable as well. Homogenization keeps the milk and cream from separating, and it is done largely because Americans have come to expect a smooth consistency in their milk, without the cream rising to the top. But homogenization causes oxidation through a slightly different process: to get that smooth texture, wholesalers pump milk under extreme pressure, which smashes the protective membrane surrounding the delicate fats in milk molecules. The process exposes the smashed, damaged fats to oxygen, resulting in oxidation. John Bunting, a dairy farmer who has been researching and writing about the milk industry for more than thirty years, told me, "Homogenization is not good . . . if you're worried about oxidized fat, it's homogenization [in milk] that's the real culprit."[6]

When used appropriately, pasteurization is vital for killing harmful pathogens, but as industrial milk purveyors use it, pasteurization also presents similar dangers to our health. Remember: any time high heat is applied to any fat, it likely results in oxidized or damaged fats. Conveniently for the industry, pasteurization means milk lasts longer on grocery store shelves. You've probably seen labels for "ultra-high temperature pasteurized" milk—this stuff will last forever, even at room temperature, but its healthfulness is highly suspect. Whole foods like milk are essentially alive, full of bacteria and enzymes that make them nutritious. What we've done to milk over the years is render it inert, making it a convenient product easily usable by the food industry. The delicate nature of many whole foods means that they are not conducive to processing and storing for long periods, something the industry has had to work around. In fact, the fats that are the most stable and least likely to oxidize with heat are the highly saturated fats we've long been told to avoid—lard, tallow, butter, and coconut and palm oils. Indeed, many processes designed to allegedly protect us—to make food last longer, less likely to spoil, or reduce our consumption of traditional fats like those found in whole milk and butter—are in fact causing damage to our heart and arteries.

It's worth understanding a bit about trans fat because its history is instructive when it comes to many other substances in our foods. Although many people take for granted that saturated fats, like butter, cause heart disease, this notion began fairly recently, promoted in the early 1950s by a biologist and pathologist named Ancel Keys. At the time, heart disease was considered an epidemic, and researchers were working to figure out its cause and ways to prevent and treat it. Keys developed the diet-heart hypothesis, also called the cholesterol hypothesis, which claimed that because elevated cholesterol in the blood appeared to be linked to heart disease, people should avoid consuming cholesterol.[7] This is where the idea of "clogging up the arteries" comes from, but in fact this is a misrepresentation. Cholesterol does not clog arteries. On the contrary, cholesterol is an antioxidant and a repair substance that appears when arteries are damaged by some other factor. When researchers find cholesterol in damaged arteries, the underlying cause is often inflammation in our bodies. What causes inflammation? Diets high in refined carbohydrates, sugar, and an overall poor diet. A poor diet also causes an imbalanced gut, which is exacerbated by the hundreds of chemical additives in our foods, some of which are proving to cause inflammation as well.

Just because cholesterol shows up at the scene of inflamed and damaged arteries does not make it the culprit. Indeed, every cell wall is made of cholesterol; it is as crucial to the survival of mammals as chlorophyll is to plants. "Why would you ever tell people to reduce their cholesterol when the backbone of every hormone that made you what you are and allows you to live day to day, is the cholesterol molecule? Every cell wall—essentially the concrete that holds cells together—is cholesterol," said David A. McCarron, a research associate at the University of California, Davis.[8] But by 1961 Keys's idea had gained significant traction: the American Heart Association (AHA) endorsed his hypothesis, and he appeared on the cover of *Time* magazine, which crowned him "Mr. Cholesterol." In the article, Keys told millions of Americans to cut back drastically on their consumption of saturated fats, like butter, cheese, and steak.[9] The AHA report endorsing Keys's cuts in saturated fats also recommended using polyunsaturated fats like those found in vegetable oils instead of saturated fats. By 1960 the organization was investing hundreds of millions of dollars in research after receiving some initial funding from Procter & Gamble in 1948. The company, along with dozens of other corporate sponsors, continues to fund the AHA to this day. When the AHA endorsed trans fats, Proctor & Gamble held the patent for the technique for partially hydrogenating oils, or making trans fats, originally from cottonseed oil, which was the basis of Crisco. Today food manufacturers continue to pay the AHA to use its Heart-Check mark on packaging for foods that are low in saturated fat and cholesterol.

With the endorsement by the AHA and *Time* magazine's cover story, many Americans switched to a diet of "heart-healthy" polyunsaturated fats and trans fats, and reduced their consumption of saturated fats—and a fad was born. Americans abandoned the long tradition of eating fats like butter or lard and instead began buying heavily marketed industrial products like margarine and vegetable oils. Both the polyunsaturated vegetable oils and artificial hydrogenated fats were a boon to the burgeoning processed food industry since they could be kept at room temperature and worked well in baked goods and fried foods. The health halo surrounding trans fats also meant that an increasing number of products could be marketed as "healthful"—and the added perk for food producers was that these oils were far less expensive than butter or lard.

And so, for the first time in history, Americans began consuming large amounts of trans fats and other vegetable fats in place of traditional fats like butterfat, beef tallow, and lard. Americans now eat more than a thousand times more soybean oil than they did in 1909. By the late 1980s these oils were in nearly every processed food and were even used by local bakeries, restaurants, school cafeterias, and any other place food was served. "These oils did not exist for the most part except in isolated parts of the world where they didn't have pigs and cows roaming around. These are not the same oils that Western societies were raised on, with the exception of olive oil," McCarron said.[10]

As most of us now know, this was a disastrous turn of events—trans fats, we have learned, are a major cause of heart disease. When Kummerow published his first paper on the harms of trans fats in 1957, he and other researchers who discovered harm from trans fats suffered marginalization and ridicule. Kummerow told me that he experienced strong and at times vicious resistance to his work and research. "Physicians believe in the cholesterol hypothesis, which says that heart disease is due to cholesterol," Kummerow said. "If you disturb anything that the physicians believe, you'll see a negative response. So you're questioning what physicians are doing and what the industry is doing, and that's a pretty powerful thing to take on." He went on: "The fat trade is a huge source of money, and so when you're telling people that what they're eating is not a good thing to eat, it also disturbs the flow of money."[11] The "fat trade" to which Kummerow was referring is the industry that produces trans fat and other sources of manufactured fats, like many processed vegetable oils. The story of trans fats and the many vegetable oils that compose so much of our modern diet—from packaged foods to restaurant foods—are the perfect illustration of how it is virtually impossible for most people to navigate the world of nutrition and health due to the deep influence of the food industry.

Fortunately, Kummerow's efforts were not in vain, although

it did take the better part of a century for everyone else to catch on.[12] Then, in 1993, Walter Willet, professor of epidemiology and nutrition at the Harvard School of Public Health, began collecting data on trans fat consumption. After eight years of studying more than 100,000 women, he found that women who ate four or more teaspoons of margarine per day had a 50 percent greater risk of heart disease than women who did not. Willet, who called for a ban on trans fats, concluded that eliminating industrially produced trans fats would prevent approximately 20 percent of avoidable coronary heart disease deaths in the United States, or as many as seven thousand deaths annually.[13]

Finally, in 2006 the FDA required that all products list the amount of trans fats on nutrition labels. However, it left a loophole: if one serving size of the food contains less than 0.5 grams of trans fat, the label can read ZERO TRANS FATS or say TRANS-FAT FREE. Confusing matters even further is that the trans fats listed are from fat sources alone, even though food additives such as emulsifiers (often listed on food packaging as mono- and diglycerides and used to improve food consistency) can also contain trans fats. But the FDA does not require that manufacturers list trans fats on their labels when they are present as food additives. This means that if you see a food label claiming it is trans-fat free, that is not necessarily the case—if the ingredient label lists partially hydrogenated oil, shortening, or mono- and diglycerides, trans fats are present. And because people often eat more than one serving size at a time, and multiple products throughout the day that contain small amounts, many Americans are still taking in significant amounts of trans fats. Even the FDA says so on its website: "If there is partially hydrogenated oil listed with the ingredients, there might be a small amount of trans fat. Selecting foods with even small amounts of trans fat can add up to a significant intake." So, despite the FDA labeling requirement, trans fats are still very much part of the American diet. They are also present in many restaurant foods because the oils are still used for frying, baking, and in salad dressings and sauces. Willet has said that "trans fats are the biggest food processing disaster in U.S. history."[14] But that remains to be seen. With thousands of additives in our foods, who knows what will turn out to be the biggest disaster?

Getting back to the breakfast example, you are likely to find questionable fats and oils in your cereals too. One of the most popular cereals in America is Post's Honey Bunches of Oats. The ingredient label lists "vegetable oil (canola or sunflower oil)." So does Kashi GoLean, another popular cereal, often considered healthy. Nearly all cereals, crackers, cookies, pretzels, breads, and other snack foods contain vegetable oils—just take a look at the ingredient lists on the food in your cabinets. "Vegetable oils" sound pretty harmless, but they're not. Vegetable oils are not stable, which means they are more likely to go rancid, or spoil. Hydrogenation solved this problem, by making them longer lasting and shelf stable, but we now know hydrogenation created dangerous trans fats. But vegetable oils are problematic for other reasons. In the 1980s scientists at the National Institutes of Health expressed concern about early clinical trials that showed subjects with diets high in soybean oil dying of cancer at alarmingly high rates. Researchers also found that people were more likely to develop gallstones if they ate diets high in vegetable oils. According to data from the U.S. Department of Agriculture, from 1912 through 2011, the annual consumption of polyunsaturated fat (mainly in the form of vegetable oils) by the average person increased from 11.3 pounds to 64.5 pounds, while saturated fat decreased from 28 pounds per capita to 13.4 pounds per capita in the same period. From packaged foods to restaurant foods to the use of vegetable oils in the home in place of butter or lard, this change represents the single biggest change in the American diet. And recall what Kummerow said: any fat that is heated becomes oxidized. Vegetable oils are heated during processing, and we have no guarantee that these oils are not oxidized by the time we eat them in our foods. Kummerow said that he never ate anything fried, used mostly butter, and consumed as little olive oil as possible, the only oil he used. In 2015, when he was one hundred years old and still researching and working in his lab, he told me that he drank whole milk every day and regularly ate pork, beef, chicken, fish, and eggs cooked in butter.[15]

Yet another problem with consuming large quantities of vegetable oils is that they contain high amounts of omega-6 fatty acids. In moderation these fatty acids are necessary and healthful, but in abundance they appear to block our absorption of omega-3 fatty acids, the important fatty acid that is lacking in most American diets. Omega-3 fatty acids, found in fish, some nuts and seeds, and pasture-raised animal products, keep our cells flexible and reduce inflammation throughout our bodies. Inflammation is the underlying factor in many conditions and diseases, including damage to our arteries. Omega-6 fatty acids, found in vegetable oils like soybean or sunflower oil, on the other hand, have been shown to promote inflammation, cell rigidity, and fat storage. Researchers have also found some evidence that omega-3s help us burn fat more efficiently.[16]

"Most scientists agree that excessive omega-6 intakes can cause problems and that omega-3 intakes are now lower than ideal in the typical American diet," the researcher Charles Benbrook told me. Researchers also believe that omega-3s and omega-6s compete for absorption in our cells, which means that the abundance of omega-6s in our diets disrupts our ability to absorb and use omega-3s. Omega-3s are crucial for eye, heart, and brain health, and there is a body of evidence that shows our current omega-6 heavy diet is taking its toll, especially on the heart. Even if you take an omega-3 fish oil supplement, as many Americans do, an excessive intake of omega-6s in your diet will render it useless.[17]

You may not think that you eat a lot of omega-6s in the form of vegetable oils, but you do. If you eat any processed foods at all, or eat at restaurants, you are taking in large amounts of omega-6s. Indeed, the main reason for the imbalance in most Americans' fatty acid profile has to do with "the huge increase in consumption of soybean and corn oil, via fried foods, condiments, and baked goods," Benbrook told me. "These common plant-based oils are very high in omega-6s and contain little omega-3s." Benbrook's main recommendation is to reduce consumption of these plant-based oils, especially corn and soybean oil.

Which brings me to yet another aspect to consider with our largely grain-based and vegetable oil–laden diets, be it our Kashi cereal or egg sandwiches: Almost all conventional crops in the United States are grown from genetically modified vegetables and seeds. Corn and soy are two of the most abundant foods in the American diet, and nearly all are genetically modified. The debate about the safety and healthfulness of genetically modified organisms (GMOs) has been one of the most contentious in the food world, with both sides prone to hysteria and hype. But several facts cannot be debated: GMOs were not introduced to the American food supply until the 1990s, so we don't know a lot about their long-term safety or healthfulness—we just don't have the data. Despite the lack of data, GMOs are now the foundation of our food supply, and nearly all of us are eating GMOs in multiple forms throughout the day. They are present in 85 percent of foods in the United States, making them nearly impossible to avoid in the typical American diet. GMO corn and soy saturate virtually every food product, including meat and dairy, because animals raised in factory farms eat GMO corn and soy.

Anything in a package that lists sugar as an ingredient comes from the genetically modified sugar beet. The vegetable oils I've been discussing—corn oil, soy oil, and canola oil—all come from GMOs. At this point, avoiding GMOs is virtually impossible unless you cook every meal from scratch, buy all organic ingredients, or patronize your local GMO-free farmer. Even organic corn is likely contaminated with GMOs because 98 percent of all corn grown in the United States is the GMO variety—and since wind inevitably carries pollen into neighboring fields, it is nearly impossible to prevent GMO corn from spreading to organic farms.

The other fact that is beyond debate from a health (and environmental) perspective is that GMOs rely heavily on pesticides and herbicides. In 1970 the biotechnology company Monsanto created Roundup, now the most widely deployed herbicide in the world, used in 130 countries. It is based on the herbicide glyphosate and is mixed with several other ingredients to make the Roundup formulation. Its use rose globally from 112.6 million pounds in 1995 to 1.65 billion pounds in 2014, and this spike coincides with the introduction of "Roundup Ready" GMO crops to our food supply, which are genetically engineered to withstand the herbicide. But all's not well with glyphosate. In 2015 the International Agency for Research on Cancer, which is part of the World Health Organization, declared glyphosate "a probable human carcinogen." According to the report, there is evidence that glyphosate causes non-Hodgkin's lymphoma, and the report also noted rare kidney and liver tumors in animals exposed to glyphosate. Other studies have found it to be an estrogenic compound that may increase the proliferation of hormone-dependent breast cancer cells. Its estrogenic properties have also shown troubling reproductive outcomes in animal studies, altering testosterone levels and testicular formation during puberty.[18] In 2016 fourteen experts published a consensus statement arguing that glyphosate is a potential endocrine-disrupting chemical that likely affects our bodies even at extremely low doses, interfering with important metabolic and hormonal processes, including increased risk for certain cancers, infertility, obesity, diabetes, and developmental problems.[19]

Because Monsanto's Roundup-Ready corn and soy have been genetically modified to withstand the herbicide, all of us are exposed to it in ever increasing amounts. The chemical can be sprayed indiscriminately, killing everything but the crops. The FDA approved Roundup-Ready soybeans in 1995, and they have been in our food supply ever since. Not that you would know—because, here again, a new food was introduced to our diets without requiring any labeling. It's also a neat financial plan for Monsanto: first, it sells its patented herbicide-resistant crops to farmers, then it sells the herbicides to be used with the crops. This resulted in a near monopoly, which is bad for farmers and bad for our economy, and the environmental consequences are profound as well. In 2013 monarch butterfly migration plunged to the lowest level in decades largely because of these changed farming practices. The explosive increase in U.S. acreage planted with genetically modified soybeans and corn engineered to withstand greater and greater amounts of herbicide has allowed American farmers to essentially wipe out the milkweed plant, which once grew between crops and is one of the butterfly's main food supplies.[20] When important pollinator species like the monarch butterfly begin to die off, it does not bode well for the resilience of our food supply or, ultimately, our survival.

One more important point about glyphosate use is that it is also sprayed all over non-GMO crops, including wheat, oats, and barley, in a process called desiccation. Desiccation is the application of the herbicide to the crop immediately before harvest to help it dry. This likely results in Roundup residue in your morning bowl of oats, cereal, or toast—even those that are labeled non-GMO. (Indeed, testing done by an independent lab found glyphosate residue on a sample of Quaker Oats Quick 1-Minute Oats that led to a lawsuit, which I discuss later in chapter 7.)[21]

Monsanto also added something to your milk without your knowledge: recombinant bovine growth somatotropin (rBST), a genetically engineered growth hormone. Dairy cows in factory farms have been injected with this hormone regularly since it was approved for use in 1993 because it boosts milk production. The hormone makes its way into our milk, ice cream, butter, and yogurt even though the FDA has never required warning labels to tell us it was entering the food stream. The growth hormone is another suspected endocrine disruptor, and studies have found links between its use and an increased risk of cancer, especially breast and prostate cancer.[22] When I asked Dr. Bruce Blumberg, a leading researcher and scientist in the field of endocrine disruptors, about rBST, he said, "We are taking in a variety of hormonally active chemicals which have the ability to disrupt our hormone signals, and for these hormonal effects, they are untested by the manufacturers or any government agency."[23] Canada, Japan, Australia, New Zealand, and the twenty-seven countries of the European Union prohibit the use of rBST.

Furthermore, the use of growth hormones means the animals are more likely to develop udder infections, leading to an even greater use of antibiotics. Because farmers routinely give their animals feed laced with low doses of antibiotics, antibiotic residue ends up in our milk and meat. As has been widely reported, antibiotic resistance in humans has become a serious problem, and consuming low levels of antibiotic residue in our foods seems to have led to a wide range of other effects—from weight gain to gastrointestinal upset.[24]

To recap: If you ate a bowl of Honey Bunches of Oats or Kashi GoLean with milk for breakfast, you ate the following: GMO vegetable oils high in omega-6 fatty acids, which are likely oxidized; GMO corn, also high in omega-6 fatty acids; GMO sugar; BHT (butylated hydroxytoluene), another potentially harmful preservative and endocrine disruptor; rBST, a suspected carcinogen and endocrine disruptor; milk protein concentrates and nonfat milk solids; and various forms of oxidized fats resulting from the homogenization and pasteurization of the milk. You probably also consumed an array of herbicide and pesticide residues from the GMO sugar, corn, and vegetable oils as well as pesticide and antibiotic residues in the milk from the cows, which were given antibiotics and fed GMO grain. And, in the case of Honey Bunches of Oats, you also consumed caramel color, which is a potential carcinogen; California requires a health hazard label for products that contain it. Contrast this with what my grandmother consumed when she ate cereal: oats and local milk from a glass bottle, all of it organic by default.

And if you think a bowl of Honey Bunches of Oats or Kashi GoLean is bad, consider another common breakfast option for busy people: Dunkin' Donuts's bacon, egg, and cheese sandwich. In this offering I counted roughly fifty ingredients, including some of the additives and preservatives I've discussed in this chapter as well as about a dozen others.* I cannot outline all the effects of those additives, but you get the point: it's not good and we have no idea—no one, however expert, has any idea—what all these additives in combination are doing in our bodies. This brings me to a crucial point: All of us are eating combinations of

---

* **Croissant:** Dough: Enriched Wheat Flour (Flour, Niacin, Reduced Iron, Thiamin Mononitrate, Riboflavin, Ascorbic Acid, Enzymes, Folic Acid), Water, Margarine [(Palm Oil, Modified Palm Oil, Canola Oil), Water, Sugar, Monoglycerides, Soy Lecithin, Potassium Sorbate (Preservative), Citric Acid, Artificial Flavor, Beta Carotene (Color), Vitamin A Palmitate, Vitamin D], Sugar, Contains 2% or less of: Yeast, Dough Conditioner (Wheat Flour, DATEM, Ascorbic Acid, Enzymes), Salt, Wheat Gluten, Dextrose, Natural and Artificial Flavor; **Glaze:** Fructose, Corn Syrup Solids, Sodium Caseinate (a milk derivative), Locust Bean Gum, Sodium Stearoyl Lactylate, Soybean and/or Canola Oil, Soy Lecithin, Sodium Bicarbonate; **Egg Patty:** Egg Whites, Egg Yolks, Soybean Oil, Water, Contains 2% or less of: Corn Starch, Salt, Natural Flavor, Xanthan Gum, Cellulose Gum, Citric Acid; **Pasteurized Process American Cheese:** American Cheese (Milk, Cheese Cultures, Salt, Enzymes), Water, Cream, Sodium Citrate, Salt, Sodium Phosphate, Sorbic Acid (Preservative), Citric Acid, Annatto and Oleoresin Paprika (Color), Acetic Acid, Sunflower or Soy Lecithin; **Bacon:** Pork, cured with: Water, Sugar, Salt, Sodium Phosphate, Smoke Flavoring, Sodium Erythorbate, Sodium Nitrite. Dunkin' Donuts website, nutrition information for "Bacon, Egg & Cheese Sandwich" on croissant, https://www.dunkindonuts.com/en/food-drinks/sandwiches-wraps/bacon-egg-cheese (accessed January 21, 2018).

potentially hundreds of different synthetically produced additives, preservatives, endocrine-disrupting chemicals, flavorings, and colorings every day. I've asked many scientists and researchers about the effects of these chemical combinations in our bodies, and the bottom line is no one knows what they are doing to us.

"We're still trying to understand the effects of single compounds before we dig into mixtures," Blumberg told me. "That's the fundamental truth. You can't understand how a mixture works unless you first understand how the individual components work." Another expert in the field of chemicals used in food packaging, Dr. Jane Muncke, said that it doesn't even make sense anymore to regulate individual chemicals because we are exposed to them in combination. "You really have to look at the mixture of these chemicals, and that becomes very, very complicated; at this point there are no real solutions for how to do that," she said.[25]

This is in part why so few studies have examined combined effects. One scientist who is studying combinations of endocrine-disrupting chemicals in particular is Andreas Kortenkamp at Brunel University in London, and his findings to date are disturbing, to say the least. In one study Kortenkamp combined a common phthalate plasticizer used in food packaging, two common fungicides used to grow food, and a pharmaceutical called finasteride, commonly used to encourage hair growth in men and to treat an enlarged prostate. All the chemicals were used at doses believed to have no adverse effects, but when researchers gave them to pregnant animals in combination, the chemicals had striking effects on the development of male sex organs, resulting in malformations in the offspring. The study authors concluded that when we ignore the combined effects of common chemicals, we are putting ourselves at great risk. Dr. Michael Hansen, a senior scientist at the Consumers Union (publisher of *Consumer Reports*), told me he wouldn't be surprised if the average American is consuming more than one hundred different additives a day.[26]

You may think that, by eating so-called healthier foods or higher-end brands like Kashi GoLean rather than Honey Bunches of Oats or a Dunkin' Donuts sandwich, you'll avoid these chemicals, but that's simply not the case. How do you avoid these chemicals? Is it even possible with our current food supply? We saw how a basic staple like milk has changed, but what about our fruits and vegetables? Is organic the answer? And does the term *organic*, which has become a stand-in for all that is natural and good, meet its high ideal? The basic nature of our simplest whole foods has changed dramatically as a result of the industrialization of our food. We're operating in a completely new landscape, one that is drastically different than it was in my grandmother's day—and now even foods with that ORGANIC label are suspect. Maybe my grandmother was on to something after all.

Losing Our Food Illusions

"Organics" and the Nutrients Missing from Our Food

As a member of the Park Slope Food Coop in Brooklyn, New York, I pay a lot of attention to what health-conscious shoppers load into their grocery carts—these are members of a progressive food co-op whose mission is, in part, to make healthier, safer foods affordable. And the co-op succeeds in doing so in many ways. But it also carries a lot of questionable items. This is partially because of the erosion of the term *organic* since the turn of the twenty-first century. Indeed, when the co-op first opened its doors in 1973, products like Annie's Macaroni & Cheese, Stonyfield YoKids Organic Yogurt Squeezers, and hundreds of other processed organic products didn't exist. Therefore you never would have found organic ingredient labels that listed substances like these (as they often do now): organic dextrose, organic maltodextrin, organic locust bean gum, organic guar gum, calcium phosphate, natural flavor, sodium alginate, silicon dioxide. This is obviously not a phenomenon exclusive to the co-op; neighborhood natural food grocers, Whole Foods, and even standard grocery stores all carry these items, and many shoppers, lulled into a false sense of safety by the term *organic*, or even *natural*, on the front of packaging, are buying foods that are no better than standard packaged products. Indeed, at the Park Slope Food Coop many people are totally convinced that they are feeding themselves and their families healthy foods. And in many cases, they're wrong.

Their misinformed buying begins with the use of *organic*, which within the last decade has become a stand-in for everything that is healthy, sustainable, and righteous. Though organic once meant something significant about how food was raised, produced, and processed, that moment has long passed. Indeed Michael Pollan—one of the best-known critics of Big Food—wrote a piece for *The New York Times Magazine* back in 2001 about the emergence of the "organic-industrial complex" and asked, "Is the word 'organic' being emptied of its meaning?"[1] I think it is safe to say that the answer to that now, more than fifteen years later, is a resounding yes. Unfortunately, it seems most Americans have not gotten this memo—even many of those loyal Park Slope Food Coop shoppers who think they are buying better foods for themselves and their families.

Many organic processed foods that I see people buy contain the ingredients I listed earlier and many other questionable additives. And nearly all these processed foods contain those ubiquitous vegetable oils that I discussed in chapter 2. Check the labels of your organic foods and you will probably find organic canola, safflower, and sunflower oil, among others. I asked Dr. Fred Kummerow, the scientist who first raised concerns about the dangers of trans fats in the late 1950s, about these organic oils and whether he thought they were any less harmful than the more common soybean or corn oil. "I don't think any oils are healthy because they are so easily oxidized. Sunflower and safflower oils are even more unsaturated than soybean, corn, or canola oil."[2] The more unsaturated the oil, the less stable and more susceptible to rancidity or oxidation it is. Bear in mind that eating a lot of processed foods, even organic processed foods, often means eating a large quantity of refined grains too—both the grains and the vegetable oils are high in omega-6 fatty acids, which appears to be problematic for our health.

But maybe you've quit the processed food world and you limit your consumption of organic processed foods, which are largely just a marketing scheme, as my grandmother suspected. What about the most basic whole foods? You shop at Whole Foods, health food stores, or the perimeter of the grocery store for whole food ingredients and buy organic milk and yogurt, and organic steaks and eggs. Unfortunately, the ORGANIC label on these staples is no guarantee that they are safe or nutritious. As is the case with organic processed foods, the label can often mean next to nothing. Take, for example, Horizon Milk, the largest producer of organic dairy products in the country—the company makes dozens of products including milk, yogurt, cheese, butter, and sour cream. Horizon organic milk, with its bright red label and happy cow on the container, gives the impression of a bucolic standard: cows grazing on acres of small-scale green pasture farms. Indeed, some of the milk that ends up in those Horizon products may well be of that nature, but much of it probably is not and is actually produced on large-scale factory farms.

The U.S. Department of Agriculture's organic standards for

dairy cattle require that cows have access to pasture for grazing at least 120 days a year, but the Cornucopia Institute, a nonprofit public interest group, found that Dean Foods (which owned Horizon and later sold it to WhiteWave Foods, which in turn was acquired by the huge multinational corporation Danone in 2017 for $12.5 billion)[3] was confining thousands of cows to large buildings and feedlots with little to no access to pasture.[4]

I asked Horizon about the size of the herds and access to pasture, and I was told that the company's "farmer partners" range in size from thirteen milking cows to twenty-four hundred milking cows, an enormous range that allows for huge differences in the way animals are treated and how milk is produced. Yet the spokesperson for Horizon told me: "We strongly believe that organic farming is scale neutral. In our view it's about how the farm is operated that matters." But scale is perhaps the major differentiator between industrial and truly organic farming. This is especially true when it comes to raising animals because part of the USDA's organic standard requires that every animal have pasture for grazing year-round. This requirement makes the practices involved in caring for a herd of fifty or even one hundred dairy cows significantly different from caring for a herd of two thousand. WhiteWave foods didn't respond to my question about how many acres each cow has access to; the spokesperson said only that third-party verifiers inspect every farm and that all are required to meet USDA standards.[5]

This brings up another troubling aspect of modern-day milk production, including the production of organic milk. Most smaller dairies can't afford to build and equip a bottling facility, which means they cannot independently produce and sell their own milk; instead they ship their milk to a larger bottling facility and are at the mercy of the practices of the larger industry. One dairy farmer in Upstate New York, Lorraine Lewandrowski, is in this position. She has a herd of sixty and raises her cattle by organic standards, although she is not certified organic (the organic certification process for some smaller farms is cost prohibitive). All her cows graze on pasture year-round and she knows each by name—in other words, her farm embodies the ideal held up by the sustainable food movement. But her grass-fed, beyond-organic milk, from cows raised using the most humane standards, is combined with milk from farms across the region, many of which are large-scale dairies that feed their cattle grain laced with antibiotics and keep the cows in confinement in factory farm settings. Lewandrowksi's milk might end up in your Fage yogurt along with the milk from dozens of other dairies. She told me that she'd love to have her own bottling plant right there on the farm, but she doesn't have the $50,000 it would cost to get it up and running.[6]

What this means for the quality of your Horizon organic milk or milk from other large dairy companies is that there are many unknowns, and as consumers we are being asked to trust a label that might not mean what we think it does. For modern-day organics this is all too often the case. Organic beef, pork, chicken, and egg production have parallels as well. Many so-called organic beef cattle are raised on feedlots where they are fed organic grain, but that says nothing about the health of the animal or the nutritional value (or lack thereof) of the meat you eat. The quality of the animal products we eat has everything to do with the quality of the food they eat; the drugs, hormones, or antibiotics that may or may not have been administered; and, some would argue, the quality of life the animal had. While there are not yet studies to prove it, common sense tells us that a stressed and sick animal is not an ideal candidate to eventually make a healthy meal. Studies do show, however, that what the animals eat has a strong bearing on the quality and nutritional value of the meat, milk, or eggs we eat.

When it comes to the quality of milk, milk that is truly organic does appear to have a nutritional edge over milk that was produced conventionally. After examining nearly four hundred samples of organic and conventional milk during an eighteen-month period, the researcher Charles Benbrook found that organic milk contained significantly more omega-3 fatty acids than conventional milk. The researchers also found that whole milk (as opposed to low-fat or fat-free versions) was even higher than conventional milk in the health-promoting omega-3s. (The organic milk that Benbrook used for this study was from suppliers for the cooperatively owned dairy company Organic Valley.)[7] And when it comes to your steak, a 2010 review of the research spanning three decades found that grass-fed beef contained higher amounts of conjugated linoleic acid (CLA) than is found in standard grain-fed beef. Multiple studies have shown that CLA protects against cancer, can lower your levels of LDL cholesterol, prevents atherosclerosis, and reduces your blood pressure. This same review also found that grass-fed or pasture-raised beef contained more vitamin A and vitamin E and more cancer-fighting antioxidants than conventionally raised beef. Furthermore, the review found that grass-fed beef has a more favorable ratio of omega-3s to omega-6s, which, as you'll recall, is a vital factor sorely missing from the American diet.[8] The research on meat, eggs, and dairy is making clear that a grass-based system for raising animals is imperative for the nutritional value of our food. But keep in mind that the food industry has managed to co-opt even the term *grass fed*. In 2015, the fast-food chain Carl's Jr. introduced a grass-fed burger on its menu, raising questions about how big companies might influence the definition of *grass fed*, much as they have influenced what is considered organic. Technically all beef cattle are grass fed at one point in their lives, but most end up in a feedlot to be fattened and grow quickly before they are slaughtered. The USDA exercises scant regulation of the term *grass fed*, and to keep up with demand some producers are claiming, with no oversight, that their beef cattle are grass fed. The American Grassfed Association, a group of producers, food industry service personnel, and consumer interest group representatives, lists producers on its website that the group has certified, which means they have met the following standards: animals are fed only grass and forage from birth until slaughter, they are not confined to feedlots, they are never given antibiotics or hormones, and all are born and raised on American farms.

There are parallels between large-scale organic or grass-fed operations and large-scale organic produce. Many of those parallels lie in scale, farming methods, and attention to the health of the soil. And there's reason to believe that the quality of organically grown vegetables can vary greatly depending on whether those vegetables were grown in nutrient-rich soil. The USDA has a certification program for organic foods, and for produce the definition states that organic farmers must "respond to site-specific conditions by integrating cultural, biological, and mechanical practices that foster cycling of resources, promote ecological balance, and conserve biodiversity." Organic produce cannot be grown from genetically engineered seed, by using ionizing radiation, or by fertilizing with sewage sludge, and the extensive list of prohibited substances includes various herbicides, pesticides, fungicides, and fertilizers. Land used for growing organic crops must be free of any prohibited substances for three years before becoming certified. As far as government oversight goes, the organic designation for produce is actually a fairly rigorous one and can be trusted to a decent extent. However, there are caveats. For example, one of the largest organic farm brands in the world is Earthbound Farm, which you have likely seen in Whole Foods and other supermarkets. Earthbound has farms from California to New Zealand to provide year-round produce from its fifty thousand acres in production. Like Horizon, it was acquired by WhiteWave Foods for $600 million in 2013 (and furthering the consolidation of industrial organic food production, you'll recall WhiteWave was then acquired by Danone in 2017). Although these farms cannot use many of the most worrisome chemicals that are applied to conventional crops, these operations grow single vast acreages of the same crop, known as mono-cropping, often without caring for the health of the soil or rotating crops. When the health of the soil is poor, the health of the plant is also poor, which means that its nutrient value may be more similar to conventionally grown crops than crops grown in mineral-rich soil that has not been depleted by a lack of biodiversity on the farm. And we have to ask: When an organic farm company becomes as big as Earthbound is, does it embody any of the original spirit or intention of organic agriculture—which has everything to do with caring for the land in a regionally appropriate way while providing local communities with safe, healthy food?

Eric Herm, a farmer and author in Ackerly, Texas, believes the difference between organic produce grown on an industrial scale and produce grown on biodiverse farms is quite significant. "What I've seen over the years is that crop rotation is not only the key to healthy soil, it is vital in the long-term health of all living creatures. There is far more microbial activity; plants are healthier and more resistant to disease, drought, and insect damage," he said. "The soil feeds the plants that feed us. Sick or weak soil will grow weaker plants with less fruit and vitality. The healthier the soil, the more vitality within the plant and the fruit it produces, therefore giving us more vitality. It's common sense, really. Organic monocropping will not have the long-term benefits of a diverse farming operation." Kira Kinney runs Evolutionary Organics, a multicrop farm in New Paltz, New York, where she farms forty acres, with twenty-five in crops and fifteen in cover crops for rotation. "I definitely think there is a difference in what I grow compared with industrial organic," she told me. "To me these two things are nothing alike. There is no holistic approach to industrial organic—it is all about yield, yield, yield. They do whatever it takes to get the most out of any given crop. Large-scale organic is much the same as conventional agriculture in that it is all numbers—get the most yield in the fewest days."[9]

Some of that common sense wisdom that farmers speak of is being replicated in the lab with findings that the fruits and vegetables we eat today are far less nutrient dense than those our grandparents ate. I would wager that while we greatly reduce our exposure to pesticides by eating organic in any form (no small benefit), the nutrient density of the produce grown on an industrial organic farm are closer to the nutrient density of crops grown on a conventional farm, likely making the following findings relevant for both conventional and industrial organic crops. In a study comparing the nutritional data for forty-three different fruits and vegetables from 1950 and 1999, researchers found significant declines in the amount of protein, calcium, phosphorus, iron, riboflavin (vitamin B_2), and vitamin C during those fifty years.[10]

You might remember hearing about a Stanford review back in 2012 that claimed to find no difference in the nutrition content of organic fruits and vegetables and their conventional counterparts. This review was hyped across major media outlets with headlines claiming no benefit from eating organic food. But the Stanford review had some significant problems. For one, the university receives large donations from Cargill, one of the largest agricultural companies in the world that uses conventional growing methods. Although Stanford officials denied any influence, the financial support of such a large player in the agricultural world could certainly have influenced the review's findings. Also curious was that both the press release and the media blitz surrounding Stanford's review downplayed two important findings: The levels of pesticide residue and antibiotic-resistant bacteria were far lower in organic produce than in conventional fruits and vegetables, crucially important aspects of organics. The review stated that pesticide residues were found in only 7 percent of organics but in 38 percent of conventional foods.[11] In a separate 2015 study, researchers found that children who ate an organic diet versus a conventional diet had significantly reduced levels of organophosphates, the most common insecticide used in agriculture, and 2,4-D, another common herbicide.[12]

In addition, Kirsten Brandt of Newcastle University published a review similar to Stanford's in 2011 and found that organics did indeed possess far greater amounts of flavanols, a type of antioxidant. Many other nutrients that Brandt analyzed in her own study and found to be greater in organics were missing from the Stanford review. "The choices they made don't seem to make sense—they seemed to include ones where the difference was smallest to begin with," Brandt told *The Huffington Post*. According to Charles Benbrook, Brandt "eviscerated their [Stanford's] methods." He told *The Huffington Post*, "The Stanford team is a bunch of doctors and clinicians, and they took on a project completely outside their training and experience. Unfortunately, their study doesn't shed any light on the subject—just a lot of smoke."[13]

The Stanford article also failed to make clear which organic crops were used in the studies under review. Were researchers analyzing Earthbound Farm's mono-cropped broccoli or the broccoli from farmers like Herm or Kinney who pay significantly more attention to the health and quality of the soil? And how long did the broccoli (for example) sit in a refrigerator before it was tested? For broccoli to maintain its nutrient density, it needs to be chilled right after harvest and eaten within two to three days. One study found that broccoli exposed to typical transport and supermarket conditions and stored for ten days lost 50 percent of its vitamin C, 75 percent of flavonoids, and 80 percent of glucosinolates.[14] Because most broccoli grown in this country comes from California, typical shipping and storage times are at least ten days for many parts of the country.[15] These are important questions and nuances often overlooked in studies and reviews like the one from Stanford.

Another factor in accounting for the loss of nutritional value is the cultivation of wild crops that led to our modern cultivated crops, which is the premise of Jo Robinson's book *Eating on the Wild Side.* Keep in mind that the size and color of fruits and vegetables matter—in many cases, we've bred much of the nutrient content out for the sake of appearance and longevity on grocery store shelves. We're clearly not going back to foraging for our own wild foods any time soon, but it is possible to determine which tomatoes, for example, are more nutritious. The redder and smaller the tomato, the higher it's likely to be in lycopene and other important nutrients. "Small, dark red tomatoes have the most lycopene per ounce, and they are also sweeter and more flavorful," Robinson writes. Which brings us to one of the most important points of all: flavor. Better flavor corresponds with nutritional value—our attraction to and enjoyment of good-tasting food is hardwired into our DNA.

Ray Bradley, a farmer in New Paltz, New York, told me that the difference in flavor in his crops, which he grows on a multicrop organic farm, and the flavor in those from a larger-scale, conventional farm that also has a stand at our local Brooklyn farmers market is drastic. "If you eat the stuff that comes off those big fields, it doesn't taste like anything," he said. Indeed, twenty of the most important flavor compounds in tomatoes all come from important nutrients, like omega-3 fatty acids and essential amino acids. That is, nutrition imparts flavor in natural whole foods.[16]

So when growers and farmers scale up, they compromise both taste and nutrition. This poses a real problem for the large-scale organic monocrop farms or big dairies that aim to ramp up production but also want to have that organic label for marketing purposes. The truth is that it is not possible to grow organic on the scale required by industrial-sized farms and still abide by true organic principles. It seems we need new terms for the various meanings that now comprise the definition of the word *organic*. Calling Earthbound Farm's broccoli organic and calling the broccoli from a local biodiverse, multicrop farmer the same thing distorts our understandings of food and nutrition—and does a disservice to the farmer who not only is producing great quality crops but is taking care of the soil and being a good steward of the land.

There is an important parallel here to the idea that "a calorie is a calorie," something the food industry has been parroting for years and is actually the basis for much of its messaging to the American people: It doesn't matter what you eat, the industry says, just don't eat too much of it. That is a convenient theory from the makers of nutrient-poor, processed foods. What "a calorie is a calorie" says is that if you go to McDonald's and get a Big Mac, fries, and a Coke for about nine hundred calories, or if you roast a chicken at home with some broccoli and a glass of milk, for a total of nine hundred calories, the effect on your body is the same. That is an absurd notion because we know that food is not just calories but a complex web of nutrients and thousands of other substances that have myriad effects on our bodies.

The ideas that we should just "count calories" or that we should just "buy organic" are related in that way. Both represent a dumbing down of complex issues that require a degree of critical thinking and understanding of nuance. The food industry is guilty of that for obvious and more dubious reasons, but the food movement is guilty of it as well. Both tend to try to simplify messages. Even Marion Nestle, a prominent food movement leader, has told me that she'd rather not complicate the message for consumers. When I asked her about the role of noncaloric factors in obesity, like the food packaging chemical bisphenol A, or BPA, for example, she said, "BPA and other such contaminating chemicals can't possibly be good for health. . . . They might have something to do with obesity—I suppose it's not impossible—but why invoke complicated explanations when the evidence for calories is so strong?" However, the most recent evidence shows that our national health crisis is a complex web of factors that cannot be boiled down to one cause, which is why it is important to understand all the issues.

So, yes, it is better to buy the Earthbound Farm's organic broccoli than the conventional broccoli—given a choice between the two, go with the organic version—but it is problematic that those are your only options. And because even the industrial organic version is costlier, it's not actually a choice for many. The third option, what some have been calling "beyond organic," is reflected in the practices of farmers like Herm, Kinney, and Bradley and is your best option. In other words, support your local and regional farmers if you can and ask them about their practices. We have to make these high-quality foods available to everyone (and later in the book I'll get to how this might be possible). The reality is that beyond-organic foods represent only a tiny fraction of our current American diet because they are not available in most grocery stores, many people don't have access to high-quality farmers' market produce or animal products, and they can be quite expensive.

Because many of us are stuck with industrial foods as the basis of our diets, our nutritional status is suffering. Many chronic diseases are the result of both the addition of problematic ingredients to our foods and a lack of adequate nutrition. With many crucial nutrients now missing from our diets, the food industry, of course, has its own solution: fortification, or the addition of synthetic versions of vitamins to our food supply. For more than half a century, two mainstays of the American diet have been fortified: milk and bread. Vitamins A and D are added to our milk, and B vitamins are added to wheat flour, which goes into nearly all bread, pasta, and cereal products. More recently, products like orange juice have been fortified with calcium or omega-3 fatty acids, and these "functional foods" are a booming sector of industrial food. But there's a problem. Although these vitamins are being added back into the food supply, many Americans are still malnourished. And contrary to what you might think—with so many Americans overweight or obese, two-thirds by recent estimates—obesity actually correlates with being malnourished. What seems like a paradox in our land of plenty, or a cruel twist, is actually a telling aspect of modern life in industrialized nations.

A 2012 study by the Harvard Medical School found that "obesity may be the new malnutrition of the homeless in the United States." Many of the foods we eat today are high in calories but low in nutrient density, which means that we can gain excess weight without contributing to the overall nourishment of our bodies. In fact, without the proper nutrients in our bodies, we can have difficulty losing weight or maintaining a healthy weight. Alternately, some nutrient-dense foods actually aid in weight loss by stimulating fat metabolism, while specific nutrients are known to do the same. Not coincidentally, many of these foods are exactly the foods we've been told to avoid for decades. For example, most Americans drink skim or low-fat milk; the two represent more than 63 percent of all dairy sales.[17] Yet you'll recall that Charles

Benbrook's study found that whole milk contains significantly higher amounts of the health-promoting omega-3 fatty acids than low-fat and skim milk—not surprising since omega-3s are indeed fatty acids. According to the study, whole organic milk contains 50 percent more omega-3 fatty acids than 2 percent milk and 66 percent more than 1 percent milk. And guess what? Omega-3s appear to play a role in fat metabolism. Perhaps that is why several recent studies have found that people who drink whole-fat milk and eat other full-fat dairy products are leaner and healthier than those who drink and eat low-fat versions. This news was reported in the press as counterintuitive. One *NPR* headline read, "The Full-Fat Paradox: Whole Milk May Keep Us Lean," but if you think back to the Whole Egg Theory, there's nothing counterintuitive at all about the nutritional superiority of eating a food in its whole form. Remember: low-fat and fat-free dairy are not whole foods.[18]

Vitamin D, another important factor in fat metabolism, is mostly acquired from sun exposure, but it is also present in certain foods, many of which we've been told to avoid because of their fat and cholesterol content: egg yolks, beef liver, butter, cheese, whole milk, and cream (vitamin D is also present in fatty fish like salmon, swordfish, tuna, and sardines). Most people associate vitamin D with milk, but that's because most commercial milk is fortified with it. The cream in whole milk contains naturally occurring levels of A and D, which was the traditional way most of us consumed enough of these important vitamins. But ever since public health officials, doctors, and dieticians told Americans to switch to low-fat dairy, and reduce butter and cheese consumption, our collective vitamin D deficit has soared.

Worldwide it is estimated that one billion people are deficient in vitamin D. In the United States, more than 75 percent of all Americans are deficient in vitamin D, even though low-fat and fat-free milk must, by law, be fortified with vitamins A and D.[19] What's wrong with this picture? Vitamins A and D are fat soluble, meaning they need fat to be absorbed by your body, which is why they come packaged with fat in whole food form. When we add synthetic A and D to nonfat or low-fat products, these added vitamins are unlikely to raise our blood levels of the vitamins because they are poorly absorbed. Synthetic fortification does a good job at preventing the severe deficiencies that result in diseases like rickets or goiter, but it doesn't meet the needs of a healthy and optimally functioning body. So while the vast majority of Americans don't have rickets—which is the result of severe vitamin D deficiency—they do have blood levels of the vitamin that are far lower than what is necessary for optimal health and chronic disease prevention. In recent years, researchers have found strong links between vitamin D deficiency and various cancers.*

This is not to say that certain supplements are not necessary and in some cases crucially important. For example, a folic acid supplement, which is a B vitamin, can prevent up to 70 percent of birth defects if taken by pregnant women. Folic acid is now added to cereals, breads, pastas, and other grain products, which typically contain little to no folic acid. These foods also happen to make up the foundation of the American diet. In reviewing official sites, including the Centers for Disease Control and Prevention,

---

* The best way to obtain optimal levels of vitamin D is an adequate amount of sun exposure—something else public health officials have told Americans to strictly avoid, much to our detriment.

the recommendation for pregnant women is to take a folic acid supplement daily and eat fortified foods, like cereal. It is only mentioned as an aside that folic acid, or folate, is naturally occurring in large amounts in leafy green vegetables, like spinach or mustard greens, as well as peas, lentils, and beans.

The New York State government website lists meal ideas for pregnant women too. To its credit, it does provide an extensive list of many whole foods in which folate is naturally present—but then, as breakfast options, the site suggests the following: two frozen Eggo waffles with syrup and fortified orange juice, or skim milk with fortified ready-to-eat cereal with a slice of wheat toast and heart-healthy spread.[20] These are the kinds of foods pregnant women—or anyone, for that matter—should avoid like the plague. This is another clear example of how the food industry has influenced dietary guidelines.

Why isn't the recommendation for pregnant women's breakfasts something like two or three eggs with one cup of spinach and a piece of whole-grain bread with butter? These whole foods would provide plenty of folate and an abundance of other nutrients, as well as satiety, thanks to the fat and protein in the eggs and butter. One cup of spinach provides 66 percent of the daily requirement for folate, meaning that if you ate another leafy green at dinner or one of the many other whole foods high in folate throughout the day, a supplement or fortified food would be unnecessary. One cup of lentils, for example, provides 90 percent of the daily recommended amount of folate, one cup of cooked asparagus provides 67 percent of the daily recommendation, and one cup of pinto beans provides 74 percent. It's easy to see that if all of us ate a diverse diet of whole foods, fortification would not be necessary. Fortification also means that the food industry can promote low-quality foods, like Eggo waffles, with the official approval of government agencies on websites and labels. Not coincidentally, these kinds of low-quality, fortified foods are also extremely inexpensive to manufacture, meaning high profit margins for food corporations. And the manufacturing of micronutrients for fortification is yet another billion-dollar industry that is part and parcel of the industrial food system. The makers of nutrients to fortify processed foods make major profits—Royal DSM, the largest nutrient manufacturer in the world, had net sales in 2016 of $7.92 billion.[21]

But fortification is a Band-Aid solution to a deeper, more profound problem. And the focus on top-down solutions, like federal requirements for fortification, distracts from the underlying issues of a nutrient-depleted processed food supply—if our foods weren't stripped of nutrition in the first place, we would not be forced to rely on fortification. As the famous food writer M. F. K. Fisher observed back in 1942, "We continue everywhere to buy the packaged monstrosities that lie, all sliced and tasteless, on the bread-counters of the nation, and then spend money and more money on pills containing the vitamins that have been removed at great cost from the wheat."[22] Fisher spotted a trend early on; it has since morphed into the billion-dollar vitamin, supplement, and functional food industries.

Even the name "functional foods" is misleading. It implies that certain foods—those enhanced by the food industry—serve some crucial function that regular foods do not. This category has an abundance of foods: orange juice with omega-3 fatty acids, yogurts with probiotics added (a comical one since yogurt already contains naturally occurring probiotics), grape juice with fiber added, to name just a few. And infant formula, which could be called the first processed food, is the fastest-growing functional food on the market, adding $5 billion in global sales in 2013 alone.[23] The global market continues to grow rapidly, with a retail value of $62.5 billion expected by 2020.[24]

The whole food matrix is not easily boiled down to a bunch of isolated vitamins and minerals, and there is something inherent in the whole food form that we cannot replicate with supplements or fortification. A 2009 review of the scientific literature in *The American Journal of Clinical Nutrition* says as much: "The concept of food synergy is based on the proposition that the interrelations between constituents in foods are significant. This significance is dependent on the balance between constituents within the food, how well the constituents survive digestion, and the extent to which they appear biologically active at the cellular level." The researchers conclude by saying, "The public may be better served by focusing on whole foods than on nutrient interpretations of them."[25]

There is also concern that the levels of synthetic vitamins added to processed foods can be harmful and even toxic. In 2004 Denmark banned cereal maker Kellogg's from adding vitamins and minerals to its cereals, saying that the company was adding toxic doses that could harm children's livers and kidneys as well as fetuses.[26] Children, babies, and fetuses are especially at risk because they have a harder time excreting excess vitamins and because so many kid and baby foods are heavily fortified. Indeed, the 1940s and 1950s saw an epidemic of hypercalcemia, or excess calcium in the blood, in children in Great Britain; the epidemic was linked to the fortification of milk with vitamin D.[27] This prompted a ban on the practice there, and many other European nations followed suit. Although some cereals in the European Union are fortified, and Finland and Sweden began fortifying their milk with vitamin D in the 1990s, many countries do not allow the practice to this day.

In the United States, a report from the Environmental Working Group found that nearly half of American children aged eight and younger consume potentially harmful amounts of vitamin A, zinc, and niacin "because of excessive food fortification, outdated nutritional labeling rules and misleading marketing tactics used by food manufacturers." All grain-based children's foods, such as cereals, crackers, and other snacks, are heavily fortified. The report found 141 overfortified products: 114 cereals were fortified with 30 percent or more of the adult daily value for vitamin A, zinc, and/or niacin, and common brands of snack bars, like Kind, Marathon, and Balance Bars, were fortified with 50 percent or more of the adult daily value for at least one of these nutrients.[28]

More is not better, especially for children and pregnant women: High doses of vitamin A can lead to liver damage, skeletal abnormalities, and hair loss, and too much vitamin A during pregnancy can cause developmental abnormalities in the fetus. Excessive zinc consumption can impair copper absorption, negatively affect red and white blood cells, and impair immune system function.[29] "Heavily fortified foods may sound like a good thing, but it when it comes to children and pregnant women, excessive exposure to high nutrient levels could actually cause short- or long-term health problems," said Renee Sharp, the Environmental Working Group's research director and coauthor of the report.

"Manufacturers use vitamin and mineral fortification to sell their products, adding amounts in excess of what people need and more than might be prudent for young children to consume."[30]

Excessive vitamin consumption may also play a role in obesity. B vitamins—the most common fortification, present in all flour-based products, including breads, pasta, crackers, pretzels, pizzas, and grain-based desserts such as pastries, muffins, and cookies—are known to promote fat gain. Baby formulas, which are also heavily supplemented, are known to significantly promote infant weight gain, especially fat mass gain, which is a known risk factor for childhood obesity. This, in conjunction with the top five food sources for the two- to eighteen-year-old age group—grain desserts, pizza, soda, yeast breads, and chicken—means that babies and children are potentially taking in far too many vitamins, which may be a factor contributing to childhood obesity.[31]

Objectively, the concept of fortification is quite odd—everything we need to eat comes from whole foods, and every component of nutrition exists therein, but our foods have been so processed, stripped, and altered that fortification has become a necessity in some cases. In specific cases fortification has been crucial to public health since food became industrialized. Diseases like rickets and pellagra have been nearly eliminated from the population in modern times, thanks to fortification. And birth defects like spina bifida and anencephaly have been reduced by as much as 31 percent and 16 percent, respectively. However, as food quality has severely deteriorated (despite fortification) in many industrialized nations in recent years, diseases like these, which were once thought to have been eradicated, are cropping up again. A 2013 study completed by the Mayo Clinic in Olmsted

County, Minnesota, found that while the incidence of rickets in children was steady from 1970 to 2000, a marked increase occurred after 2000. Rickets is a disease in which bones don't mineralize properly because of a lack of vitamin D, calcium, or phosphate. Rickets has also seen a recent resurgence in Britain, where doctors attribute it to lack of sunshine and playing outdoors as well as poor diet.[32]

It's clear that we have put ourselves in a rather precarious situation: On the one hand, our foods have become so depleted that it seems reasonable to fortify them to ensure we are not missing out on important nutrients. On the other hand, fortification presents its own set of problems, which we may only be beginning to understand. It seems obvious that the solution is a return to a diverse, whole foods, beyond-organic diet—but most of us don't have access to that anymore. Even those of us who think we eat healthy foods are clearly at risk: we need to think twice about those Annie's Organic Cheddar Bunnies or those supposedly healthy fortified cereals covered with Horizon Organic Milk—and even our organic steaks or Earthbound Farm's organic broccoli.

Industrializing our food system has resulted in complex and confounding problems—from the chemicals in our food packaging, to our everyday exposures to myriad additives, to the lack of nutrition in the foods we grow and eat—our quest for a truly healthy diet is nowhere near simple or easy to understand. And, as I am about to show, the effects on our bodies also go beyond the immediately obvious; indeed, the industrial food system is quite literally changing our bodies.

PART TWO

How the Industrial Food System Is Changing Us from the Inside Out

Our foods have changed drastically—that much is clear. But what is perhaps most shocking is the fact that this is rapidly changing our bodies. Our ultraprocessed diets, beginning with infant formula, baby and kid foods, and then the foods we continue to eat as adults, are literally changing us from the inside out. The billion-dollar food industry, which essentially has free rein in how it markets its products, has developed niche markets for baby foods and children's foods—new and completely unnecessary inventions. These products have undermined traditional knowledge about what to feed babies and children and are leading to dire health consequences.

When a woman is pregnant, what she eats while her fetus is developing is crucial for her child's lifelong health. Then, whether or not she breast-feeds also sets the stage for the child's future health and susceptibility to disease. Breast milk is imperative for optimizing a baby's health on many levels. The flavors transferred to the baby in the mother's breast milk (ideally from a diverse, whole foods diet) appear to play a major role in whether the baby will be open to eating a nutritious diet throughout life. Equally important, breast milk is essential for establishing a healthy microbiota, the collection of bacterial cells that reside in and on us—something we've discovered only recently. And our processed food supply, beginning with infant formula, is at the root of our inability to transfer a healthy microbiota to the next generation.

As adults our reliance on a diet of processed food is compromising our microbiota and putting us at risk for a long list of chronic and degenerative diseases—from gastrointestinal problems to mood and brain disorders. The industrial food system is changing aspects of our bodies in unprecedented ways—from shaping food preferences in the womb to the extinction of certain bacterial species that have been in our bodies since the dawn of our existence.

The Industrial Food Setup

Getting Us (and Our Children) Hooked from Day One

Grocery stores everywhere are home to entire sections devoted to baby and children's food. The shelves are lined with jars, cereals, crackers, snack packs, infant formula, and brightly colored plastic squeeze packs of fruit and vegetable mixes. That might not strike you as odd—after all, it has become a common sight in grocery stores—but it is. Of course, many of these items are labeled as organic and natural, a huge selling point, especially for baby and children's food. But it's odd precisely because there's nothing natural about the idea of baby food at all. The very notion of baby or kid food sprung from the industrialization of our food, just like the organic and natural categories did. And just as everything in my grandmother's day was organic and natural by default, everything she and her family ate at the time was also baby food. That is, the food the parents ate was also the food babies ate. Until quite recently in our eating history, no one wondered what to feed babies and children—babies drank breast milk until they were weaned and ate what the adults ate, often mashed, pureed, or prechewed. In a preindustrial food world, fewer options meant less confusion—babies and children ate nutrient-dense whole foods just as their parents did, and that was true only seventy-five years ago. The idea that babies and kids need some kind of specially manufactured formula or food is a creation of the food industry in its quest to continually carve out niche markets and novel food products. And these markets are booming. Baby formula is the fastest-growing "functional food" on the market, and the combined sales of baby food and baby formula reached $30 billion globally in 2015.[1]

Manufactured specialty foods have become such an accepted part of raising babies and children that the concept is taken for granted—but this is actually an aberration. Like so much else about our current food system, it represents a deep rupture in our understanding about food, nutrition, and even our survival as a species. The popular notion today is that babies and children prefer bland white foods—think rice cereal—but how did we come to accept this? It has no historical precedent or scientific basis, yet we take it as a fundamental truth. How often have you heard well-meaning parents say things like "My kid will eat only white bread, chicken fingers, pizza, peanut butter and jelly, or mac and cheese"? We never stop to consider why these children are such picky eaters; we simply accept that they are and cater to them. After all, that list reads like any kids' menu in any restaurant across the country. So it's hardly surprising that this is what children (especially in the United States and other industrialized nations) want to eat—it's what the industry and their parents have been feeding them all along.

It even begins in the womb. Researchers at the Monell Chemical Senses Center, a nonprofit research organization in Philadelphia, have made important discoveries about the role a pregnant woman's diet plays in developing her baby's future taste preferences. They found that flavors in the amniotic fluid and then in breast milk transfer to the baby in an education process that tells the baby what foods are safe and preferred by her mother. This research has profound implications for how our diets change and shape the lifelong food preferences of future generations. According to the Monell researchers, babies who were exposed to a wide range of flavors in the womb, through their mothers' diverse and varied diets, and later through flavors in breast milk, are more open to different and varied flavors as babies and children. Conversely, babies who were not exposed to those varied flavors show aversions to trying new and different flavors. "It's our fundamental belief that, during evolution, we as humans are exposed to flavors both in utero and via mother's milk that are signals of things that will be in our diets as we grow up and learn about what flavors are acceptable based on those experiences," Gary Beauchamp, the director of the Monell Center, told me. "Infants exposed to a variety of flavors in infancy are more willing to accept a variety of flavors, including flavors that are associated with various vegetables and so forth and that might lead to a more healthy eating style later on."[2]

This means what the mother eats while pregnant and breastfeeding is crucial. It also means that babies fed formula are at a disadvantage. Because the flavors in formula never change—unlike breast milk, which is constantly changing based on what the mother eats—babies fed formula are less likely to eat a wide and diverse range of foods and instead prefer the monotonous, or "one-note," flavors commonly found in processed foods. "We learn about food long before our first taste of food. A fundamental feature of mammals is the first way we learn is through the flavors that transfer to amniotic fluid and then mother's milk," said Julie Mennella, a biopsychologist and researcher at the Monell Center. "Babies whose mothers eat certain foods are more accepting of those foods at weaning. There are also differences between breast-fed and formula-fed infants . . . because these babies have gotten more cues [by] learning about these flavors in mother's milk. So it's a beautiful, elegant, simple system, and then baby learns what mother is eating, what she has access to, and what she likes; this way Mom teaches us what foods are safe." Mennella emphasizes that this system works only when the mother eats a wide range of nutritious foods so the baby is exposed to a range of flavors.[3]

After weaning, the baby should be eating the same foods as the rest of the family—not baby foods. Mennella laments the introduction of bland baby cereals and other bland baby foods. "We can't just feed babies separately from their families or caregivers. Like with all cultures, eat the healthy foods: fruits and vegetables, grains, and meats that you like because it gives the baby the opportunity to like them as well," she said. In the case of many baby food products, like bland cereals, these are often foods that are never consumed again. "You don't have to learn to like no-texture food, like pudding or cereal; but you do have to learn to like the complex variety of foods we have." Babies are also born with an innate preference for sweet foods, but they need to learn to like other flavors and eventually accept them with repeated exposure. The trouble is, in our current food landscape, many babies are never getting that exposure because their mothers are simply not eating those foods, nor are they feeding those foods to their children. "Children are very vulnerable to the current food environment because their biology favors sweet and salt—but they need experiences to learn to like these other foods," Mennella said. "It's going to be harder for them to accept vegetables than fruits. Baby doesn't have to learn anything when it comes to sweet, they are born liking sweet. They've got that in their DNA."[4] Babies' preference for sweet is a consequence of evolutionary selection for favoring nutrient-rich mother's milk and, later, energy-dense fruits. It also steers them to avoid bitter, poisonous plants.[5]

Flavor preferences beyond our innate preference for sweet are highly elastic and vary greatly across cultures and time. The key is exposure. Consider how children in other cultures eat the foods they are offered. This was the subject of a *New York Times Magazine* piece in 2014:

> Breakfast for a child in Burkina Faso, for example, might well include millet-seed porridge; in Japan, rice and a putrid soybean goop known as *natto*; in Jamaica, a mush of plantains or peanuts or cornmeal; in New Zealand, toast covered with Vegemite, a salty paste made of brewer's yeast; and in China, *jook*, a rice gruel topped with pickled tofu, strings of dried meat or egg. In Cuba, Brazil and elsewhere in Latin America, it is not uncommon to find very young children sipping coffee with milk in the mornings. In Pakistan, kids often take their milk with *Rooh Afza*, a bright red syrup made from fruits, flowers and herbs. Swedish *filmjolk* is one of dozens of iterations of soured milk found on breakfast tables across Europe, Asia, the

Middle East and Africa. For a child in southern India, the day might start with a steamed cake made from fermented lentils and rice called *idli*.[6]

One hallmark of the processed foods many Western children begin their palate education with is their relatively uniform texture, with added artificial flavors. These foods also tend to be high in sodium, poor-quality liquid fats (vegetable oils), and sugar, creating a highly palatable texture, flavor, and mouth feel. Whole fresh food's, on the other hand, can be pungent, tart, and bitter, with a variety of different textures. Researchers at the Monell Center have identified sensitive time periods during which babies develop taste preferences, and one of those is before they are three and a half months old. Therefore, if a baby misses out on this crucial window of taste preference development, it may be hard to change his preferences later.

In a separate study researchers found that the food industry works to "fundamentally change children's taste palates to increase their liking of highly processed and less nutritious foods."[7] Bettina Cornwell, an author of this study and professor of marketing at the University of Oregon, told me that the food industry is unquestionably shaping the palates of children, often before they have a chance to learn to like nutritious whole foods. Her studies show that the key is repeated exposure. "Children can learn to like healthy food, just the same as they can learn to like junk food," she said. This is why she does not buy the idea of the "picky child eater" who will eat only a limited number of foods. "There are real and true clinical instances of neophobia [the irrational fear of anything new]—but this does not describe the majority of children and majority of the resistance to nutritious, healthy foods," she said. "You have parents who say, 'My child only eats peanut butter and jelly; my child only eats chicken nuggets' . . . and then they say, 'My child is a picky eater' because the child now won't accept steamed Brussels sprouts. But has the child had adequate exposure to those foods? That is the question."[8]

In another major study, researchers found that a child's junk food diet, as opposed to a traditional or health-conscious diet, was positively associated with being a difficult eater. This suggests that exposure to junk food actually creates the picky eater, not the other way around, as many American parents commonly believe. The researchers write, "The results do allow an interpretation that junk food in the diet may change attitudes toward other foods."[9]

In her studies at the Monell Center, Mennella has found that it takes on average ten to twelve offerings and tastings of new foods for babies and children to accept them. Babies will often make faces, but that doesn't mean they will reject the food; getting the baby to accept the food requires the parent to persevere. "It's really important to teach moms that Baby will make that face, but don't give up. It takes a lot longer for the face to change than for intake to change," Mennella said. Both she and Cornwell emphasize the importance of teaching babies and children to like nutritious foods as early as possible. "We need to get them off to an early start," Mennella said. "It's easier to introduce foods when the baby is younger—in the first few years of life. And what they are eating at this time is the best predicator of what they are eating at school age." Cornwell agrees that the most important time to intervene is early in life. "We should focus our research, our policy changes, and our federal, state, and local funding on getting children off to the right start," she said. "Behavior change is so much more difficult—when you're already overweight, or when you already have these food patterns . . . we can't change those behaviors as easily as we can grow up with the right ones. That's why I orient to the youngest of children."[10]

It is also why food marketing orients much of its advertising to the youngest children; like researchers, marketers know that teaching young children to like fast food and packaged foods early on will influence their lifelong dietary choices. "You can imagine that going to McDonald's, three or four times, six or eight times, twelve or eighteen times, starts to set a pattern where those foods become accepted," Cornwell said. In one fascinating study, Cornwell and a colleague looked at the toys that accompany fast-food meals for children, specifically toys that were part of a set. When food manufacturers create toys that are part of a set, it encourages children to nag their parents to return to the restaurant to complete the set, thereby ensuring that the child is getting repeated exposure to these foods. "Very young children are cognizant of set completion," Cornwell said. "And they hold set completion goals—so this toy, for example, is part of a set of three, so you have a fast-food meal with one toy, and you become aware that the other two toys exist, and then you want to return to get the second in the set of three, the third in the set of three—it is this concept of repeated exposure."[11] The toys that are part of a set with fast-food meals become powerful influencers, even of how the food tastes. In her study Cornwell found that when the third toy in the set was presented as part of the healthy meal, the children found that the meal tasted better.[12]

This makes the marketing of children's food particularly troubling. While it is widely accepted that the marketing of junk foods encourages kids to eat more of those foods, leading to obesity, diabetes, and even heart disease in children—not something to take lightly—there is something even more insidious at work if these foods are creating various feedback loops that result in children wanting to eat more of these same foods, become desensitized to particular flavors and components in whole foods, and significantly alter their taste preferences.

Likewise, the marketing of infant formula is especially worrisome because of its profound implications for long-term health. With the knowledge that what a baby is fed from day one will have consequences for his lifelong food preferences and long-term health, the current methods of marketing infant formula appear almost criminal. Consider the labels on infant formulas, which are intended to convince new parents, who are already quite vulnerable, that formula will ensure a healthy baby. The label on Enfamil's newborn formula says, "DHA is a fatty acid that acts as a building block for a baby's brain. And Enfamil Newborn has this brain-nourishing nutrient that is important to a baby's development." Enfamil also says that its formula is "inspired by the changing nutrition of breast milk," and it contains "choline and DHA, important brain-nourishing nutrients." The label further states that "Enfamil's nutrition helps support early milestones like focusing and smiling." Another formula, Similac, claims it is "closer than ever to breast milk" and contains "DHA for brain, Lutein for eyes, and vitamin E for development." Good Start's formula label declares that it is "modeled after the complete nutrition of breast milk" and has a "unique blend of DHA prebiotics and probiotics." These claims mean little to nothing, and no regulatory agency checks them for accuracy. Like much else in our food supply, infant formulas are only minimally regulated by the U.S. Food and Drug Administration (FDA).

Marsha Walker, executive director of the National Alliance for Breastfeeding Advocacy, said that the FDA does not test any formula to see if its manufacturer's claims are accurate. "All the FDA does is require that there are twenty-nine specific ingredients in any standard infant formula that is placed on the American market. The agency accepts assurances from formula companies that their products are GRAS (generally recognized as safe), but it does no testing to see if this is true.* The FDA is not funded to actually test any formula to see if it does what it claims to do," Walker said.[13] She added that no infant formula is close to breast milk in its composition, efficacy, or health outcomes, and for manufacturers to claim this on labels is deceptive.

Kathy Dettwyler, a professor of anthropology at the University of Delaware and an expert on breast-feeding, said that while the formula makers can do a satisfactory job of matching basic components like fats, carbs, and sugars, they will never be able to replicate breast milk. "For the last couple of decades the research was showing that the long chain omega-3 fatty acids, DHA and ARA, were important for brain development and were found only in breast milk," she said. "Which, at first, the formula companies were insisting didn't matter; but then they gave in and all started adding DHA and ARA to the formula. But those are just two of who-knows-how-many different components—hundreds, thou-

* "Generally recognized as safe" actually means little, which I will explore in more detail in chapter 7.

sands of different components found in breast milk that aren't in formula. When they say formula is close to breast milk, it's like saying Erie, Pennsylvania, is closer to Hawaii than New York City is—yes, it's true, but it's still not close. I don't think they'll ever be able to replicate what's in breast milk or the process of breast-feeding."[14]

Breast milk is a miraculous substance. It contains all the nutrition a newborn baby needs to survive. For at least the first six months of life, a baby needs only mother's milk—she doesn't even need water. In a continual process, a mother is literally dissolving her own body into the milk to nourish her baby. Her body accesses the fat in her hips and butt that it has been storing throughout her life, releasing those brain health–promoting omega-3 fatty acids. And that's just one example of the countless vitamins, minerals, and other fatty acids the baby receives from breast milk for optimal health. Breast milk is also highly protective, conveying customized immune support in a manner that is almost unfathomable. When a mother breast-feeds, the saliva in her baby's mouth enters the mother's nipple, allowing the mother's body to sense what the baby needs in terms of germ- or bacteria-fighting antibodies. Her breast milk is then customized to target an infection. Mothers also gather bacteria from their babies' bodies when they touch, kiss, and hold them, and their breast milk responds with antibodies their babies need. The mother also gives her baby antibodies from illnesses that she has already had and fought off, hopefully sparing her baby. Even while breast-feeding, if a mother gets the flu, she will pass antibodies on to her baby through breast milk, preventing the baby from becoming sick. Breast milk is therefore constantly changing based on the needs of the baby; researchers refer to it as a dynamic, bioactive fluid that changes in composition, varies within feeds, and changes throughout the day and between mothers.[15] Breast milk is also teaming with protective bacteria, as well as the specific components that feed the bacteria; these will colonize the baby and become his microbiota, which has profound consequences for his lifelong health (I explore this further in chapter 5). With these kinds of endless variations and combinations of breast milk ingredients custom made for each baby as needed, manufacturers will never be able to truly replicate mother's milk. After all, this intricate system of feeding babies has been honed over more than two hundred million years of mammalian evolution. This means that formula-fed infants' long-term health is being compromised on three fronts: they are missing out on the crucial nutrients found in breast milk, they are deprived of customized immune system support and a healthy microbiota, and they are potentially being set up for a lifelong preference for processed foods, thereby exacerbating the other deficiencies.

But formula makers and processed food producers are solely looking to increase their bottom line, so much of this science is ignored or dismissed. In fact, formula makers aggressively seek out new consumers by recruiting mothers in delivery wards and providing them with free goody bags full of formula samples and other swag. This practice was largely unquestioned for decades, but recently many U.S. hospitals have banned the goody bags in recognition of the importance of breast-feeding. These bans were instituted after studies found that women given these samples were less likely to breast-feed or breast-fed for less time than mothers who did not receive the samples. The studies also found that mothers perceived handouts as a medical seal of approval. Today

67 percent of top hospitals no longer distribute formula company–sponsored discharge bags, formula samples, or formula company promotional materials to mothers in the hospitals' maternity units.[16] Still, the use of formula is exceedingly common.

It's worth remembering that infant formula, like baby food, wasn't even available until fairly recently and its use is quite new. Before the early 1900s, mothers who couldn't breast-feed for medical reasons didn't have the option of infant formula and often used wet nurses, or women who were employed to breast-feed children other than their own. As the practice of wet nursing faded with an increased focus on sanitation, mothers started preparing their own foods, known as "dry-nursing." Doctors at the time noted a failure to thrive in many babies who were dry-nursed—and, no wonder, because this substitute was usually just a mixture of water and commercial flour or skim milk and water. Decades earlier, in 1867, the famed scientist Justus von Liebig had developed the world's first commercial infant formula, which eventually led to the development of many others that would soon capture a huge market, like Nestlé's infant formula. As it happens, Liebig, who was an organic chemist, is also known as the father of the fertilizer industry, and many of his inventions helped to usher in modern agriculture and modern food systems. And doctors first prescribed commercial infant formula at about the time that the first industrial foods were making their way into the home in the early 1900s. But because infant formula was expensive, most mothers still breast-fed, a trend that continued into the 1950s. By the late 1950s, however, commercial formula was readily available and less expensive—and breast-feeding was falling out of favor, with only 30 percent of mothers reporting that they breast-fed their week-old babies in 1958.[17] This trend marked a shift in the cultural attitude about the science of nutrition and health. Modern advances in chemistry and science seemed like potential cure-alls while baby-feeding traditions seemed antiquated and old-fashioned. Formula feeding was a trend that even my grandmother wasn't immune to, despite her rejection of many other industrial food products. "All the doctors were telling us to use formula then, so we listened—what did we know?" she told me.

There was also a sense that breast-feeding was somehow a primitive practice, writes the food historian Amy Bentley. "There was something backward, or even distasteful, about breastfeeding one's infant in mid-twentieth-century America; exposed breasts and suckling children elicited too much discomfort and were too reminiscent of the dark-skinned women from developing countries displayed in full color in the pages of *National Geographic*."[18] Despite its now well-known health benefits, breast-feeding has never really recovered from these shifts in cultural attitudes. One woman I was counseling on nutrition told me that "breast-feeding is not for everyone," when I inquired about her newborn baby. This struck me as incredibly odd—the idea that such a natural and biological process as breast-feeding could become so foreign to us as a culture that we could just choose to reject it on a whim.

This is partly a result of policies and workplaces that do not make it economically feasible for women to stay home and nurse their babies. With no federal policy in place for paid maternal or parental leave, work culture in the United States makes it exceedingly difficult for women to breast-feed, even when they want to. According to the Centers for Disease Control and Prevention, in 2016 the percentage of women who exclusively breast-fed six months after giving birth was an abysmal 22.3 percent. The percentage of women who do some breast-feeding at twelve months was 30.7 percent. The American Academy of Pediatrics recommends that women breast-feed for at least twelve months and until the child is two years of age, if possible. The World Health Organization's recommendation is that women breast-feed for two years or more. With six months as the absolute minimum for the health of both the baby and mother, a federal law guaranteeing six months of paid maternal or parental leave would go a long way toward improving public health in this country. And while you wouldn't know it from the statistics on breast-feeding, the pendulum has indeed swung back the other way: all the major health organizations and government agencies endorse breast-feeding. That's because the science on the health benefits of breast-feeding is irrefutable. Well-established health benefits of breast-feeding for the baby include decreased risk of asthma, childhood leukemia, childhood obesity, ear infections, eczema, diarrhea and vomiting, sudden infant death syndrome, and type 2 diabetes.[19] Less known, though, is that whether a baby is breast-fed appears to have lifelong effects on her food preferences throughout childhood and into adulthood, pointing to yet another clue to how and why we've become a nation of people so dependent on highly processed foods.

Another strike against formula feeding is that these infants are also significantly more likely to be obese and overweight as children and adults. A baby who is breast-fed for nine months is 30 percent less likely to be obese than a baby who is never breast-fed. And studies suggest that this protection against obesity persists into teenage years and adulthood. Even controlling for lifestyle factors such as parental weight, dietary factors, physical activity, socioeconomic status, and parental education levels, the protection against obesity for breast-fed babies remains. This is likely due to a combination of myriad factors related to the content of breast milk as well as the presence of many substances in breast milk that are still unknown, although studies are pointing to some explanations. Formula-fed infants have higher levels of insulin in their blood and a more prolonged insulin response to feeding than breast-fed babies. Excess insulin circulating in the blood is not good—it leads to the storage of greater amounts of fatty tissue in the body; eventually too much insulin circulating in the body leads to diabetes. The high amount of protein in formula may also lead to increased amounts of insulin in formula-fed babies.[20]

Breast-fed babies, on the other hand, have higher concentrations of leptin circulating in their bodies; leptin is a hormone that controls appetite and regulates the healthy storage of body fat. And researchers are also finding that obesity does more than put children at risk for all its related diseases—there is evidence that obese children are less sensitive to flavors, which perhaps closes the loop between our dietary habits and our food choices. In a German study of two hundred children, obese children rated all flavor concentrations lower than their normal weight peers on a flavor intensity scale. The obese children also had greater difficulty identifying salty, bitter, umami, and sweet intensity than their normal weight peers. This desensitization may be because leptin, in addition to regulating appetite and making us feel full, seems to affect the sensitivity of our taste buds. Obese or overweight people may be resistant to leptin, making them feel hungry and driving them to eat more.[21] And because obesity in toddlerhood and childhood is strongly correlated with obesity throughout life, the change in taste perception probably persists as well.

In a related study, Julie Mennella and her colleagues at the Monell Center found that sweet flavors, which can act as an analgesic, were not as effective in obese children. For this study Mennella reported that babies will cry less and leave their hand in a cold-water bath longer with sweet flavors in their mouths. But for obese babies and children, the sweet flavor did not work as well as an analgesic.[22] "I hypothesized maybe it's because of some disruption in the opioid system, so maybe they need more sweet to get the same effect," she said.[23] The opioid system is a powerful pathway in the brain that controls pain, reward, and addictive behavior. Opioids, a variety of compounds that act on specific receptors in the nervous system, are present in foods and drugs and can also be produced within the body. In a study published in *The American Journal of Clinical Nutrition*, researchers found that obese adults were less able to detect the presence of fatty acids than adults of normal weight.[24] This could mean that a mechanism for recognizing how many fatty acids we consume with our foods has been fundamentally disrupted. What all these findings point to is rather profound: the processed food industry is subverting a delicate and finely tuned system, honed over millennia to attract us to the foods that we need to be optimally healthy.

Consider these other compelling findings, which dovetail with Mennella's hypothesis about the opioid system. In a 2013 study, scientists looked at the food preferences in the offspring of rats born to mothers who ate a junk food diet. The researchers found that when they fed pregnant and lactating mothers foods like Cheetos, Nutella, Teddy Grahams, and Froot Loops, their offspring showed a desensitized reward pathway in the brain. "The best way to think about how having a desensitized reward pathway would affect you is to use the analogy of somebody who is addicted to drugs," Jessica R. Gugusheff, lead author of the study, told me. "When someone is addicted to drugs they become less sensitive to the effects of that drug, so they have to increase the dose to get the same high. In a similar way, by having a desensitized reward pathway, offspring exposed to junk food before birth have to eat more junk food to get the same good feelings." Gugusheff explained that highly palatable junk foods light up the mesolimbic reward system in the brain along the same pathways as drug use.[25]

These findings all point to the conclusion that the industrial food system and its big players are doing much more than making us overweight. They are fundamentally changing our bodies and how our bodies respond to these new food creations. But these changes are obscured by the alleged controversy and conflicting information about what comprises a healthy diet for babies, children, and adults alike—indeed, these controversies are the life blood of the food industry. And the industry has confused and tricked even the most highly educated and savvy among us. The road to our current food and health crises is paved with profit for the food industry, largely stoked by controversies in nutrition science and a general ignorance about what actually constitutes healthy food and a healthy diet. This confusion often comes from the industries themselves, seeding doubt in the minds of consumers about what's healthy. The industry also rejects the science and frames what we eat as a matter of personal choice. In doing so, it keeps important social programs, as well as governmental regulation, out of the conversation. The narrative that the individual has the right to choose what to eat free from any kind of regulation or oversight placed on the industries overshadows the real health ramifications of those choices.

And the debate about the importance of breast-feeding is much like the debates about our eating habits. Oddly, despite all the research showing the overwhelming benefit of breast-feeding for both babies and mothers, the controversy about feeding formula versus breast-feeding persists—even among highly educated parents, doctors, and researchers. Many women advocate for formula, insisting that women who advocate breast-feeding have become self-righteous "lactavists," as one writer puts it.[26] The formula advocates end up repeating the mantra of the food industry, that every woman's choice should be respected, especially the choice to feed formula. But I wonder: If women really knew the truth about all the benefits that breast-feeding conveys, would they see it as a choice? Shaming "lactavists" represents a backlash against a backlash of sorts, with women countering the narrative that "natural is best" and instead defending large corporate interests like formula manufacturers and Big Food companies.

Indeed, the idea that infant formula is a perfectly fine stand-in for mother's milk, or that consuming processed baby and kids' foods fortified with synthetic vitamins is the same as eating nutrient-dense whole foods, is part of the industrial food system's mythology that has persisted for almost a century. It stems from the concept that whatever nutrients we need can be isolated in a lab and pumped into processed food products with no ill effect. It completely ignores the reality of the Whole Egg Theory and is part of the misguided notion that "a calorie is a calorie." When it comes to mothers and babies, for decades the medical establishment paid little attention to the quality of the food—so long as the mother was gaining weight and the baby was growing, and continued to thrive after birth, all was thought to be well. Which, if you think about it, makes perfect intuitive sense so long as you are eating actual whole foods. But with the industrialization of our food supply, all the rules have changed. Recall the New York State government's food recommendations for mothers-to-be, which include empty calories like Eggo waffles and ready-to-eat cereal or, worse, the harmful fats found in the so-called heart-healthy spreads. As a result, many babies are getting processed foods in the womb and then, with the introduction of infant formula, as their first food at birth. Then babies and children are immersed in a processed food culture that has completely normalized the nutrient-poor foods they are expected to accept and like. Those bland white foods made from refined and processed grains, added sugar, and poor-quality fats in the form of vegetable oils—products like bread, pasta, crackers, and cereals—industrial foods that deprive children of nutrients critical to their mental and physical development.

Indeed, the processed foods many of us are now encouraged to eat are damaging our ability to taste foods and perceive their many flavors—and may even result in the rejection of the wide variety of flavors and textures found in nutritious whole foods. And although formula makers were not initially aware of this phenomenon, they also benefit when children begin eating solid foods. Nestlé, one of the largest producers of infant formula, also has a significant market share of processed food products; the company profits from the sale of formula, which then helps shape palates so that the child prefers industrially processed foods later—perhaps for life.

Newborn babies and young children are the most vulnerable among us, and the food industry is capitalizing on that by pushing formula on new mothers and later luring young children to eat its foods with ploys like the promise of a toy set. The innate wisdom that has been honed over millennia by cuing us into what foods will keep us optimally healthy is being sabotaged by an industry solely focused on profits. In short, Big Food is setting us up for lifelong health problems and a lower quality of life—and in the case of young people today, the most dire ramifications are coming to fruition: they will have an even shorter life span than their parents.

Living in a Microbial World

Industrially Processed Food and Our Guts

The University of California, Davis, employs the world's leading veterinarian for foaling race horses, John Madigan. Mares move into the foaling barn in late winter or early spring, and they deliver at about the same time with vets on hand to help. In March 2016 the first foal born came down with diarrhea and died the next day. More foals were born in the days that followed, and they too were falling ill. After several days Madigan realized that the cause was the bacteria *Clostridium difficile*, or *C. diff.* for short. You have probably heard that it is notoriously hard to treat and often antibiotic resistant. It's also a nightmare for hospitals because it is spread by bacterial spores and is highly contagious. Worried that all the foals would succumb, Madigan decided to tap some of the resources at UC Davis. He sought help from Bruce German, a chemist and professor of food science and technology who has done extensive research on human babies and breast milk. German's innovative work on the microbiota—or the trillions of bacterial cells that live in and on us and are vital to our health—has shown that a compromised microbiota probably sets the stage for a wide range of ailments and diseases, from allergies to cancer. After about ten days German and his team had a plan to help Madigan and the sick foals: They developed a bacterial cocktail to give to the animals at birth. By this time fourteen or fifteen foals had been born, and all were sick. But Madigan got to the sixteenth newborn foal, gave it the bacterial cocktail—comprised in part of a crucial strain called *Bifidobacterium longum subspecies infantis* (*B. infantis* or *Bifidobacterium* for short)—and the horse escaped the infection. He gave it to the next two foals to be born and, again, no problems.[1] "The bacteria eradicated the problem," German said. "We would never imagine exposing human babies to *C. diff.* to test whether a bacteria would work [to counter *C. diff.*], and in fact we probably wouldn't even do it in animals, but this just happened and it was so instructive."[2] The beneficial bacteria they gave the foals overwhelmed and crowded out the bad bacteria, rendering them harmless in the gut.

The foals were susceptible to *C. diff.* because of the ubiquitous use of antibiotics, sometimes for good reason—but antibiotics kill commensal bacteria, or those that protect against pathogens and normally transfer from mother to her offspring. In this case antibiotic use among the mares resulted in foals that were "horrifically unprotected" at birth, German told me. In humans the overuse of antibiotics is backfiring too. The ramifications go beyond the widespread and well-documented cases of antibiotic resistance to more subtle and lesser-known effects on our microbiota. (The Centers for Disease Control and Prevention estimate that one in three antibiotic prescriptions is unnecessary. That's forty-seven million prescriptions each year.)[3]

The microbiota is essentially an organ with vital functions for our health. In an adult its collection of trillions of bacterial cells weighs in at about three pounds and can be acquired only through our exposure to bacteria—specifically, through vaginal childbirth and breast milk. That is, it's not a given like our heart, liver, lungs, or brain. Research suggests that when we subvert the processes of childbirth and breast-feeding, we sabotage its development. Because the microbiota has been measured and studied only recently, we have missed its vital role in our long-term health for several generations. Many autoimmune conditions, like food allergies, environmental allergies, asthma, and atopies of the skin—which have risen steeply in recent decades—likely originate in the first year or two of life with an overactive immune system that has not been properly tuned by a protective microbiota.

A compromised microbiota also seems to promote inflammation, which begins in the gut and then spreads throughout the body, whether in the arteries or heart, the joints, or the brain. These discoveries thoroughly change the way we view diet-related diseases like diabetes, heart disease, arthritis, cancers, and even neurodegenerative diseases like autism and Alzheimer's—as at least partially linked to an inappropriate inflammatory response that began in the gut in infancy.

How does the microbiota tune the immune system and prevent inflammation? It's an important question, one that researchers are working to answer. Inflammation, the primary response of our immune system, has two types, acute and chronic. Acute inflammation is what happens when your body is responding to an injury, something like a sprained ankle, broken bone, bacterial infection, or a virus. This is a temporary life-saving response, summoning immune cells, nutrients, and hormones to help heal whatever has gone wrong. On the other hand, chronic inflammation is the result of an immune system that has been responding inappropriately for a long time. In the latter case, the body perceives something as a threat that really should not be. Common examples include environmental allergies to substances like pollen or allergies to foods like nuts or fish. The problem with a chronic immune response is that the immune cells begin attacking our tissues, cells, and even our organs. This manifests in a range of diseases common in the Western world. Chronic inflammation is linked to osteoarthritis, inflammatory bowel disease, Crohn's disease, diabetes, heart disease, and cancer, among many others. So, the question is, Why do so many of us now have diseases that seem to be caused by chronic inflammation? The answer appears to come back to our processed, nutrient-poor diets and the microbiota.

Inflammation and immunity are intricately linked. Living things have essentially two types of immunity, innate immunity and acquired immunity. Innate immunity is what we are born with and what gives us the ability to recognize and fight off harmful exposures. The central molecule involved in innate immunity is endotoxin, not only in humans and animals but also in plants. Endotoxins are part of what's known as gram-negative bacteria, or pathogens such as *salmonella.* When exposed, our immune system mounts a response to these endotoxins, primarily through inflammation, and in ideal circumstances wipes them out. But if a baby doesn't have the right balance of good bacteria to crowd out the endotoxins, they can cause a lot of problems—some fatal and others leading to chronic conditions and disease. In premature babies, the presence of endotoxins can be deadly, as is seen in cases of necrotizing enterocolitis, which causes extreme damage to the intestinal tract. It is most common in premature babies who are fed formula. Hospitals now strongly advise mothers of premature infants to give their babies breast milk—often pumping it and leaving it with nurses in the neonatal intensive care units—since it is so important in saving these babies' lives. In fact, Mark Underwood, a pediatrician at UC Davis Children's Hospital and a colleague of German's, is utilizing their findings on the important role of *B. infantis* by feeding premature babies a combination of the bacteria and breast milk and seeing good results.[4] But even in full-term babies, German believes the immune response in those who are fed formula or even breast milk without being inoculated with *B. infantis* can manifest in inflammation and may be at the root of colic, allergies, asthma, and even obesity, diabetes, heart disease, and cancers later in life.[5]

The second type of immunity, acquired immunity, develops through our exposure to pathogens and antibody systems, largely inherited from our mothers at birth and through breast-feeding, as well as from the environment around us. But the body must make important decisions about what is a threat. As a result, as a baby is gradually exposed to all kinds of benign substances, her immune system should recognize that these are not threats and build up tolerances to them. But if the baby tries her first peanut butter, for example, and coincidentally has a bloom of endotoxin in her gut, her immune system is going to respond by associating the molecules in peanut butter with an inflammatory response and will inappropriately respond to it as a dangerous substance. Therefore the first couple years of life appear to be key to educating and establishing a healthy immune system. In the case of the ever-increasing rise in food and environmental allergies, the body is having an overblown inflammatory response to what should be perceived as benign molecules.[6]

Because we have ignored the importance of the bacterial transfer from mother to baby during the vaginal birth process and during breast-feeding and have indiscriminately used antibiotics for decades, we have caused one crucially important and protective strain of bacteria normally found in the baby's gut to go extinct in the Western world. A woman of child-bearing age who was born by C-section, fed formula, or received antibiotics at any point in her life—or if this is true of her mother or grandmother—does not have the important bacterial species *B. infantis.* Let me reiterate this point because it is so important: if *any one of those scenarios* applies to you, your mother, or your grandmother—which is nearly everyone in the Western world—you no longer harbor *B. infantis* in your body and are therefore unable to pass it on to your children. "We've realized just in the last five years that antibiotics, infant formula, C-section, and ruthless hygiene have basically taken these bacteria out of the Western world, and actually most babies don't get it," German said. He then added this crucial point: "The demise of the bacterium coincides most eerily with the perplexing increase in all of these diseases of autoimmune origin, like atopic dermatitis, food allergies, environmental allergies, colic, asthma, et cetera, et cetera." Indeed, atopic dermatitis in babies born between 1960 and 2000 has risen fivefold, and type 1 diabetes incidence in children has also increased fivefold.[7]

According to German and his team's research, without *B. infantis*, a baby's microbiota is compromised from day one with lifelong ill effects. "The central benefits of having a microbiota dominated by *B. infantis* is that it crowds all the other guys out and especially the gram-negative bacteria," German said. Therefore, the development of our immune system, which begins immediately at birth, is entirely dependent on the components of breast milk and the transfer of *B. infantis* from mother to baby, in German's view. "I think this is what evolution invested in—making sure that endotoxin wouldn't bloom in a baby," he said.[8]

But we've missed this for a really long time. "The truth is, the world should have understood the importance of breast milk a long time ago," German said. "We study mothers and babies and breast milk for a lot of reasons. One of them is that they are the most important subjects for relevant health research, full stop." He refers to breast milk as the Rosetta Stone for nutrition—it provides the blueprint for what human beings need not only to survive but to grow and thrive in sometimes harsh conditions. The biological process of lactation, which has been shaped by two hundred million years of evolution, provides "a complete, comprehensive, nourishing, protective, and preventative diet to mammalian infants," German said. And, he added, since mother's milk is (or, rather, was) the only thing that infant mammals ate, its safety was ruthlessly tested throughout those two hundred million years. German emphasized how lactation is therefore *the model* for how to nourish humans completely, comprehensively, preventatively, and safely, as well as sustainably and deliciously.[9]

I was confused when I first heard about *B. infantis* for two reasons. First, much of the research on the microbiota suggests that a healthy microbiota is a diverse one—meaning that the more bacterial strains and species you have in your gut, the better. But what German and his team found is that a healthy baby's gut is populated almost exclusively by *B. infantis*; it dominates his entire lower intestine, crowding out nearly all the other bacterial strains. Second, I mistakenly assumed that someone who eats a healthy whole foods diet, especially one full of fermented foods like yogurt or sauerkraut, would have a gut populated with all the beneficial bacteria a person needs. But because *B. infantis* has been rendered extinct in the Western world, the only way to get it, or pass it on, is to be inoculated with the specific strain.

Was German saying that women in the Western world—even those who eat a nutritious whole foods diet, give birth to their babies vaginally, and breast-feed exclusively—are now incapable of giving their babies these all-important bacteria? "That's exactly right," he said. "In order to have a particular organism in an ecosystem, and we'll take the intestine as an example, you have to be explicitly inoculated with at least a few of those bacteria alive, so they can grow; and if you don't acquire those by inoculation, they can't grow." He used the analogy of a rose garden to illustrate this point: "If you want to have a rose garden in your backyard and you never plant roses, you can put all the water, fertilizer, and sunlight into that garden you want, but you are going to be very frustrated with the lack of roses."

Indeed, according to German's research, 97 percent of American babies do not have *B. infantis* in their gut. On the other hand, the majority of infants in less-industrialized countries have a gut dominated by *Bifidobacteria*. Research across countries worldwide has shown that children without *bifidobacteria* are six times more likely to have allergies and type 1 diabetes, and they tend to have an unhealthy body weight.[10] And in a cohort of poor women in Bangladesh, babies with *B. infantis* had better responses to vaccines two years later than babies without it. (Even babies born by C-section in Bangladesh were colonized with the bacteria, which German attributes to a less sanitized environment and the constant close contact between babies, their mothers, and other relatives early on).[11] An added benefit, not only for babies but for new parents, is that when *B. infantis* flourishes in the baby's gut, the baby has fewer and better formed (read less messy) stool. German and his team found that babies who were inoculated with the bacteria and fed breast milk went from six messy diapers a day down to one diaper. What does this say about gut health? "The immune system is working better, there is less inflammation, there is better integrity in barrier function [the lining of the gut], which means less irritation by pathogens and endotoxins in the first year of life," German said.[12] Not only does this mean a less fussy, healthier, and more robust baby, but down the line the implications are profound for preventing chronic diseases like cancer, especially those in the gastrointestinal tract. Bear in mind that the United States has seen an alarming rise in the incidence of colon and rectal cancers in people in their twenties and thirties, something previously unheard of.[13]

Thinking back to those foals: rather than treat the *C. diff.* infection with antibiotics (which obviously doesn't work well, since it is an antibiotic-resistant pathogen), the answer was adding protective bacteria, which crowd out and overwhelm the bad bacteria, rendering them harmless. What if we've been approaching the problem of antibiotic resistance from the wrong angle? What if, rather than developing antibiotics, which kill not only harmful but also beneficial bacteria, we instead added beneficial bacteria that crowd out the bad? That would solve the problem without creating new ones. This represents a paradigm shift for medicine. And it has striking parallels to our modern agricultural systems.

Industrial agriculture has been hell-bent on elimination: killing pests, fungus, weeds, diseases—you name it and we have developed a toxin to kill it in an effort to keep the crops we want alive. But we've seen the consequences of this kind of system—it is not sustainable and it creates its own set of problems: resistance to these chemicals puts farmers on a "pesticide treadmill" as more and stronger chemicals are needed to kill off resistant pests, weeds, or diseases. The medical paradigm of treating infection is not much different: identify the pathogenic bacteria and kill them off. Yes, in many cases antibiotics are a modern miracle that spare us from life-threatening infections. However, the emergence of superbugs like *C. diff.* that are resistant to antibiotics present a looming global health crisis. But perhaps even more damning is the discovery that we have eradicated strains of beneficial bacteria that have been with us for all of human existence. Not only do we have an immediate public health crisis, but we have set the stage for long-term, chronic diseases that we are only beginning to understand—especially because the loss of these beneficial bacteria often results in health problems that don't manifest for years, decades, or even several generations. German believes that the absence of *B. infantis* in the Western world over the last several generations could well be at the root of many of our modern health crises.

German and his colleagues provide an alternate model: provide the health-promoting bacteria and allow them to do their job by crowding out and overwhelming the bad bacteria; it works with the natural systems of the body, not against them. This is evolution's smart answer to a world teeming with all kinds of microbes: rather than kill off the bad, set the stage for conditions in which the good bacteria multiply, overwhelming the bad and rendering them harmless in their diminished numbers. But the right conditions are key. We need to be inoculated with the right bacteria and then feed those specific bacteria the right food. What do the *B. infantis* bacteria need to eat in order to proliferate and protect the baby? Not surprisingly, the bacteria need breast milk. Indeed, German learned about *B. infantis* in the first place by studying the components of breast milk; to his surprise it contains an abundance of oligosaccharides, a type of carbohydrate that babies are incapable of digesting. It is the third most abundant component of mother's milk, and he wondered: Why would a process that has been perfected over the course of evolution provide something in abundance that a baby could not digest? After all, the mother is literally dissolving herself to make food for the infant, which means that nothing in her milk is inefficient or evolution would have driven it out. Then German realized that these carbohydrates were not feeding the baby—they were feeding *B. infantis*. "There are more indigestible oligosaccharides in breast milk than there is protein—so basically it is *as important to feed the bacteria* as it is to feed the babies. I am still astonished by that," he said.

This is such a key insight about what we have lost as a result of our modern birth and feeding methods. In the best-case scenario, a healthy baby's gut is populated almost exclusively by *B. infantis*, and the oligosaccharides continue to feed the bacteria, to the exclusion of nearly everything else, until the baby is completely weaned off breast milk, which ideally is around two years of age and when the baby's immune system has developed properly.

Infant formula, on the other hand, does not contain the complex oligosaccharides needed to feed *B. infantis*. Formula companies will often put a few strains of *Bifidobacterium* into formula, but these are strains found in adults, not the strain that is essential for babies, which is highly sensitive to temperature and oxygen—and not conducive to sitting on a grocery store shelf for months at a time. Although formula makers would like to say that their product is similar to mother's milk in this respect, it simply is not. "The term *Bifidobacterium* is like the term *dogs*," German said. "We understand that there is considerable functional diversity among dogs, and if we are looking to tow a dog sled, all breeds need not apply. So, if Danone et [alia] say that they are adding *Bifidobacterium* to infant formula; you really have to ask the question: Is it the *Bifidobacterium* that inhabit breast-fed infants' intestine? Unfortunately, not."

But don't forget—most babies in the Western world don't have *B. infantis* in the first place, so the oligosaccharides in breast milk can't do the job of feeding the bacteria. The consequences of this extinction appear to be profound.

If we knew that C-sections and infant formula compromised the development of the heart or the lungs, would we be so quick to encourage them? And what about our antibiotic overuse? Just like the foals whose mothers got too many antibiotics, human babies are directly exposed to large amounts—beginning on the day of birth, when a baby is supposed to receive a transfer of protective bacteria from his mother. A baby gets antibiotics on day one for some common reasons. First, when a baby is born by C-section, the mother is often given intravenous antibiotics to prevent infection, which changes her bacterial population and the baby's too. Second, women in the United States who test positive for Group B streptococcus are given intravenous antibiotics so the baby won't acquire strep during labor and delivery. But it is not clear that antibiotics are warranted, because only one in two hundred babies gets sick from the bacteria, which means 199 other babies are treated with antibiotics unnecessarily. In addition, it is not entirely clear that giving antibiotics to mothers with Group B strep even prevents infection in the newborn.[14] Many other countries do not practice this routinely, and they do not have higher rates of babies' falling ill from the bacteria. All babies born in the United States also get an antibiotic eye ointment immediately following birth to prevent infections in case the mother has gonorrhea (even when she has already tested negative for it). In many of these cases, we are giving millions of babies' antibiotics when they are not clearly warranted.[15]

The consequences of compromising the baby's microbiota are only beginning to be understood, but it is becoming increasingly clear that the risks to the baby's lifelong health may outweigh any immediate risks or the justification for antibiotic use. In the case of C-sections, much of the exposure to antibiotics could be reduced by eliminating these often unnecessary surgical procedures. Of course, in some instances they are necessary to save the life of the mother, baby, or both, and we are lucky to have a safe option in those cases. However, they are often performed without medical necessity, either because pregnant women schedule them for convenience, or by doctors who want to rush the birth process or are not trained well enough to help a woman give birth naturally. In the United States, one in every three births is now a C-section. Since 1996 the use of C-sections has increased 50 percent, and if this trend continues, half of all babies—two million a year—could be born surgically by 2020.

Because traveling down the birth canal is *the critical means for acquiring your microbiota*, those who miss out on this process face lifelong health consequences.* In multiple studies babies born by C-section harbored no vaginal bacteria and instead had strains of bacteria found on the skin—in some cases these were bacteria present on a delivery room nurse or doctor, not the mother's bacteria. And the development of the baby's intestine with normal and healthy bacterial colonization was delayed at one month, two years, and even seven years after C-section birth.[16]

Once the baby is born, the instinct to latch on to the mother's breast ensures the further transfer of bacteria from mother to baby, first in the colostrum and then in breast milk. The baby first gets the crucial bacteria both in and on his mouth from his mother's vaginal and digestive tracts during labor, and then the colostrum and breast milk feed these particular strains in a perfectly orchestrated combination that allows those beneficial and protective

* Some women who have caesarean sections are considering seeding their baby with their vaginal bacteria. This new practice is still largely unknown in the medical community. Erica Sonnenburg, a researcher who studies the microbiota at Stanford, told me that when she had her daughter by C-section, she swabbed her baby with her vaginal bacteria in the hope of transferring some of her bacteria to the baby. This process involves placing a cotton swab in the vagina before the C-section and then swabbing the baby's body, face, and mouth. More research is needed to determine whether this process colonizes the baby with the mother's bacteria. (Erica Sonnenburg, interview with author, January 15, 2016, Stanford, California.)

strains to flourish in his gut. It is difficult to overstate just how important these processes are to the health of the baby—not only immediately but for life. That's because the infant microbiota educates the immune system and programs the metabolic system, and what we now are beginning to understand is that the disruption to the natural processes of forming a protective microbiota seems to have disastrous consequences leading to the development of autoimmune diseases and conditions as well as metabolic dysfunction.[17]

But we have completely ignored this for decades, moving toward not only an industrialized birth method but an industrialized way of feeding babies—infant formula. Infant formula—the first processed food—undermines the life-promoting benefits of mother's milk. Formula is a poor substitute for breast milk nutritionally, and it deprives babies of their ability to develop a healthy and robust microbiota. German believes we have missed this for a simple reason: research and medical science has focused on the diseases of middle-aged white men and has largely ignored mothers and babies. The United States has invested billions of dollars researching treatments for male health issues—high cholesterol, high blood pressure, heart attacks—instead of focusing on prevention. German's model regards the exchange between the mother-infant dyad as the foundation for lifelong health. Of course, had we recognized this, or if we finally do recognize this, the chronic diseases of middle-aged men could disappear, or at least greatly diminish.[18]

The continued excessive use of antibiotics throughout infancy and childhood also wreaks havoc on the microbiota—the average

American child receives one course of antibiotics every year until age eighteen.[19] Thus it appears that the combination of C-sections, ubiquitous antibiotic use, infant formula, and ruthless sanitation, as German puts it, has had some unforeseen and rather frightening repercussions for our health, with significant cost to the health of babies born in our modern sterile environments.

Our Second Brain

You know those butterflies you get before a date or big interview? Or that sinking feeling in your gut that alerts you when something is not quite right? These are examples of what researchers are now calling our "second brain," or the gut-brain axis, to describe the intricate connections between the gut and the brain. We now know that the gut and the brain communicate in many ways that we often don't recognize at all. First, the brain and the gut communicate through the vagus nerve (there are actually two, but they are referred to in the singular), which runs from the brain, down the esophagus, through the heart, into the abdomen, and all the way to the transverse colon. The microbiota can activate the vagus nerve and strongly affect the brain and mood, potentially leading to disorders like anxiety and depression and even affecting cognitive function.[20] The gut also communicates quite extensively with the brain through the enteric nervous system, which is a mesh-like system of neurons that lines and controls the function of the gastrointestinal system.

At its most basic, the brain-gut axis is something we have all also experienced through pain. A stomachache after eating certain foods and the discomfort of gas and bloating are examples of bacteria in the gut that are responding to whatever foods you've just sent their way. The relationship between the microbes in our gut and our immune system, which also communicates extensively with the brain, is part of our ancestral biology. "The brain is not just this ball of fat in the head, it is this whole network that connects all around your body, and, not surprisingly, we've got a whole lot of nervous tissue all around our gut, and we are paying close attention neurologically and immunologically to the bacteria in our gut," German said. Immune cells line our gut and monitor the bacteria there, so information flows constantly back and forth between the gut and the brain. "We are living in a microbial world," German said. "Their job is to eat us, our job is to try and keep that from happening, so you would imagine investing in a lot [of] surveillance at the very least, and you would also imagine that variations and alterations in that microbial population could trigger immunological activation, pain, inflammation, neurological stress, and mood swings, et cetera—it just makes such perfect sense."[21]

This means the health of your microbiota probably affects your brain health throughout your life. In fact, the modern alterations to the microbiota could also play a role in the soaring rates of neurodegenerative disorders like autism, attention deficit/hyperactivity disorder (ADHD), and Alzheimer's disease. The scientific literature has documented that children diagnosed with autism spectrum disorder (ASD) have a greater incidence of gastrointestinal disorders and distress than the general population. These children also tend to have highly restrictive diets and will only eat a limited variety of foods, with strong aversions to many foods. One autistic child I knew would eat only white rice, white breads, sweet foods, and milk. She refused to even try fruits, vegetables, whole grains, and an array of other whole foods. Researchers have been asking why this is the case for some time now but are only beginning to think about autism in relation to the microbiota and gut health.

Dr. Richard Frye, director of Autism Research and the Autism Multispecialty Clinic and codirector of the Neurometabolic Clinic at Arkansas Children's Hospital, is researching the connection between the health of the microbiota and children with autism. "For some kids, we definitely see major changes in the diet really affecting behavior. And we believe there are several pathways that can affect children with autism, one of them being the microbiome," Frye said. One of Frye's findings is that children with autism tend to produce too much propionic acid, a naturally occurring acid in the body produced by a particular strain of gut bacteria. In a healthy system the right balance of this acid is a tool for the mitochondria (the powerhouse of each and every cell in our bodies), but too much of the acid can cause mitochondrial dysfunction, which usually means that the cell has difficulty fully using food and oxygen to generate energy, resulting in an array of health problems. Why do these children have too much propionic acid? Frye said it could be because some kids have an overrepresentation of *Clostridium*, a bacteria that can make propionic acid. But why? It could certainly be the diet, Frye said. When your diet has too many refined carbohydrates, like bread, pizza, pasta, cookies, or crackers, you aren't producing many short-chain fatty acids, which are crucially important for keeping inflammation at bay throughout the body.[22]

In Frye's clinic doctors often recommend a modified Atkin's Diet, or a low-carbohydrate diet, which seems to help kids with autism as well as kids who experience a lot of seizures. They've seen that a low-carb diet improves mitochondrial function in some children, meaning their cells are better able to function and carry out their duties. "Why exactly that is, is not clear but a low-carb diet is thought to be anti-inflammatory in some ways, but maybe it also changes the microbiome—shifts it in some way that improves physiology," Frye said.

Yet another connection to the microbiota that many autism researchers and doctors have identified is that certain nutrients, in particular folate and vitamin B_{12} levels, are consistently low in the brains of children with autism. When Frye gives children with autism injections of B_{12}, symptoms significantly improve in some cases. Oral doses of B_{12} don't seem to work as well, perhaps an indication that because the microbiota is involved in the metabolism of folate, a dysfunctional gut could be preventing the vitamin from reaching the bloodstream and then the brain.[23]

Again, the mother's being deficient in nutrients seems to have lifelong effects for her children. It is well known that folate is a crucial nutrient for preventing severe birth defects, such as neural tube defects. Although the synthetic fortification of foods may prevent the most obvious birth defects, it may not be as effective as eating foods with naturally occurring folate for preventing subtler changes to the developing baby. Folate is found in leafy green vegetables, beans, peas, and lentils, among other whole foods, but many people aren't eating these foods anymore and instead rely on fortified processed foods and supplements. I asked Frye if he envisions the creation of an "anti-autism" diet for prospective parents to adhere to. "Yes, I think so. But it's not just autism—along with increases in autism, we're also seeing other neurological disorders, including ADHD," he said. "And [we're] also [seeing] lots of allergies and other types of immune disorders, so I wouldn't pigeonhole it to just autism. I think it would be important to intervene in that way to improve the outcome of child health in general."[24]

The mother's health, both what she is exposed to and what her diet consists of, is vital to the health of her baby. "A really important point is that it may not be the kids themselves, but we are finding that many things that happen prenatally in moms is very, very important," Frye said. "We talk about all these environmental influences, but the most important environment that anybody is ever going to be exposed to is their mother's womb, where anything that happens—inflammation, dietary restrictions—can cause amazing problems on a growing fetus, so that is a whole other level of things we still really need to look at." In his clinic Frye often sees children "after the car wreck," when so many things have already gone wrong for the child. "The problem in kids with autism, even when you get their diet under control, you still see these abnormalities. They don't go away magically, and it may be that the problem is we're looking at things so far down the line. We may have gotten into a vicious spiral, and it may be very hard to correct," he said.[25]

This is a crucial, more subtle point often missing from the charged rhetoric around autism, its causes, and its treatment. The key seems to be prevention in the form of the mother's health both before and during pregnancy, with particular attention to her nutrition status, her diet, and the state of her microbiota and her immune system. We must also consider our exposure to environmental chemicals—unlike generations past. Once the developing fetus has undergone certain changes, whether because of inadequate nutrition or exposure to some environmental chemical, correcting them can be difficult, if not impossible. Therefore, even extreme dietary changes, supplementation, or other treatments for autistic children can prove futile—which is devastating to parents who often pin their hope on anecdotal stories or protocols endorsed by celebrities.

In fact some research seems to indicate that the state of the mother's immune system is *the determining factor* in whether her child will be autistic. At this point German is hesitant to make any definitive statements about the role of the missing *B. infantis*, or the state of the microbiota and its relationship to ASD, but he is a part of a large multidisciplinary team of scientists and researchers at UC Davis that is trying to unravel the complex science of the microbiota and what this could mean for the development of autism. German echoes Frye's belief that preventing autism in the first place is the best way to deal with this epidemic. But it appears that addressing the immune system of the mother could be the key. And since the immune system is so intricately connected to the microbiota and vice versa, autism probably has microbiological factors. More specifically, if the mother's gut is in a dysbiotic state (meaning she has an imbalanced microbiota), her baby is also likely to be dysbiotic. "You can see how the microbiome, even if it is not causal, is part of the overall problem, because autistic kids seem to be unusually neurologically sensitive to when they have an intestinal upset," German said. "That doesn't mean that the bacteria are causing autism, but they are making the child behaviorally disruptive."[26]

The Adult Microbiota

A lifelong diet of processed foods and a lack of nutrients and their effects on the gut may also be contributing to the brain dysfunctions we associate with aging, such as Alzheimer's and Parkinson's. These changes may be mediated through a dysbiotic gut, or exposures to environmental chemicals and various food additives may act directly on our brains—or both. Some researchers have begun calling Alzheimer's type 3 diabetes since Alzheimer's appears to be a form of type 2 diabetes that affects the brain. What's more, a person with type 2 diabetes has more than double the risk of developing Alzheimer's than a person without the disease.

Diabetes causes insulin-resistance in our cells, and when the cells in the brain become insulin resistant, a person starts to develop some of the symptoms associated with Alzheimer's, such as memory loss and personality changes. In fact, insulin resistance both impairs cognition and seems to be implicated in the formation of plaques in the brain, called beta amyloid plaques, that are associated with Alzheimer's disease. Other studies have shown that diets high in carbohydrates while low in fat and protein may increase the risk of mild cognitive impairment in the elderly, even without diabetes. Despite much evidence to the contrary, doctors are still recommending a low-fat diet to people diagnosed with prediabetes or diabetes, when in fact they should be lowering their intake of sugars and carbohydrates, not fat or protein.

Indeed, our Western diet is changing the adult microbiota in a way similar to what German found in the baby's gut. The researchers Justin Sonnenburg and Erica Sonnenburg at Stanford University have shown that many important species in the adult microbiota are extinct or going extinct in the Western world. Just as the loss of bacteria in the infant's microbiota means that even the right oligosaccharides from the mother's milk cannot feed it, once any strain of bacteria is eliminated from a population, even eating the right foods cannot restore the lost species to the microbiota. The Sonnenburgs found that our drastic shift to an industrial food diet is severely lacking in what feeds the bacteria in our gut.

One of those components is commonly called fiber—and the per capita consumption of fiber in industrialized nations is extremely low, about 15 grams per day. This is about one-tenth the amount of what human beings ate for millennia as hunter-gatherers and rural agrarians. As a result, the bacteria in our gut have been drastically reduced during the course of several generations, a direct result of our modern largely fiber-free, hyperprocessed, overly sterile diets. The Sonnenburgs believe that we are "starving our microbial selves," which is the title of their 2014 paper that outlines the extinction of bacteria still present in hunter-gatherer populations and rural agrarian populations but absent from the guts of people living in the industrialized world.[27]

FIBER

The term *fiber* is a gross oversimplification of the highly complex substrates that feed the bacteria in our guts. In fact researchers have not even identified the multitudes of carbohydrates that feed various bacteria. Erica and Justin Sonnenburg call these substrates "microbiota-accessible carbohydrates," or MACs, the carbohydrate foods our microbiota is capable of using. MACs are far more complex than the familiar notion that there are only two types of fiber, soluble and insoluble. The microbiota is an ecosystem that functions based on what it is fed and not fed, and it is all about the incredibly complex and still mostly unknown multitudes of carbohydrates. "Some bacteria eat some carbs and not others, and right now, we can't measure it well," Bruce German said.

Erica Sonnenburg told me about the Hadza people in Tanzania whom she and her team are studying. This is a group of roughly two hundred to three hundred people who still live and eat exclusively as hunter-gatherers—they are one of the last groups of people to live like this on the planet. They are living like our ancestors lived for most of the past 150,000 years or so, and they are living in the geographical location that was the cradle of human evolution in East Africa. Therefore, the Sonnenburgs believe this group is our best approximation of what our ancestral microbiota looked like for nearly all of our evolution—and the Hadza are consuming 100 to 150 grams of fiber per day, or ten times what most Americans are consuming. "Dietary fiber and the complex carbohydrates found within them is the major currency within the gut, and if you remove ten times that amount of food, then a lot of species of microbes just die off," Sonnenburg said. As we have seen, other factors are also involved in the die-off of microbes in the Western microbiota, including antibiotic use, the cleaning up of our water supply, advanced sanitation, the increased use of C-sections, and formula feeding—but for adults, diet is the most crucial aspect in determining the health (or lack thereof) of the microbiota, Sonnnenburg said.[28]

It gets more complicated still when considering what the Sonnenburgs and a team of researchers found in a 2016 study of four generations of mice.[29] The researchers compared the microbes in the gut of two sets of mice. They gave both groups of mice healthy microbes from a human donor and then fed one group of mice a diet high in fiber. They fed the other group a low-fiber diet but kept the number of calories, from protein and fat, the same as the first group's. Within two weeks the researchers saw extreme differences in the two sets of mice. More than half of all the microbial species declined by 75 percent in the guts of the group fed the low-fiber diet, and many species disappeared altogether. After seven weeks the researchers put the low-fiber group of mice back on a high-fiber diet and found that fully one-third of the bacterial species *never recovered*, meaning that the loss of bacteria was permanent.

What's more, this dwindling of bacterial species persists and worsens over generations. Each successive generation had fewer bacterial species than the one before. By the fourth generation, nearly three-quarters of the microbial species present in the great-grandparents' guts were absent in their own. "Our multigenerational experiment shows that diet could be a major factor in why these bacteria are not present in the Western gut," Erica Sonnenburg said. "As we shifted to an industrial, processed food diet, there wasn't the dietary fiber there that these microbes require."[30]

And here's the crucial point: The absence of dietary fiber for most people in the Western world during the last seventy-five to one hundred years means that our microbiotas have also changed rapidly—the trouble is, our human genome has not changed along with it. The human genome has adapted and evolved over hundreds of thousands of years, but the microbiota adapts incredibly rapidly. "Your microbiota can change in a week if you change your diet," Sonnenburg said. "So the concern is, because the Western diet is relatively new and has had a dramatic effect on our microbiota, that our human genome has not had the time to adapt to the community that we now harbor in our gut. If you have a microbiota composition that is relatively new, compared to your human genome, there may be important interactions that have been lost, and now that community isn't playing so nice with our human side."

What happens, exactly, when the microbiota and the human cells "don't play so nice"? At this point, there are many unknowns, but one thing that researchers do know is that when the microbes are starved of the food they need to live, they begin to eat the mucosal lining of the intestine, since it contains carbohydrates. When the microbes begin to eat the mucosal lining of your intestine, the barrier between the bacterial species and your human cells begins to thin; when the human cells of the intestine sense that bacteria are encroaching, they mount an immune response to prevent the bacteria from getting into your bloodstream, resulting in chronic inflammation—that is, your immune system is on constant alert and the tissues throughout your body are permanently inflamed. "This can result in cell damage because a pro-inflammatory state is meant to ward off infections, but it is not meant to be there chronically, and that's the beginning of inflammatory bowel disease and colon cancer—and this amount of inflammation never dampens down because those cells never feel like they are at a safe distance from the bacteria, so they are always working to kill those microbes," Sonnenburg said. "The human body wasn't set up for that type of scenario; it was set up to fight off acute infection, not constantly fight off these bacteria in our gut."

As I mentioned, this also points to why colon and rectal cancers are now on the rise in people in their twenties and thirties in the Western world—something that was practically unheard of even a decade ago. People born in 1990 have double the risk of colon cancer and quadruple the risk of rectal cancer than people born in 1950.[31]

Although the gut is not meant to be porous to microbes, it is meant to be porous to the fermentation by-products of the microbes. These by-products are like the waste, or feces, of the microbial cells, and they consist of thousands upon thousands of chemical molecules used by our bodies—the vast majority of which have yet to be studied. The most studied by-products are the short-chain fatty acids, which are known to dampen the immune response in the host, preventing a range of diseases like inflammatory bowel disease, colon cancer, diabetes, and heart disease. Recall that Frye believes the absence of these fatty acids may be part of the autism puzzle as well. Short-chain fatty acids are present in your body only if you eat the types of foods that produce them—high-fiber foods like dark leafy greens, vegetables, beans and legumes, and whole grains.[32] On the other hand, if you eat a diet high in refined carbohydrates like white bread, pasta, crackers, and cookies, these rarely break down into short-chain fatty acids.

It's important to remember that you first must have microbes that are capable of feeding on the short-chain fatty acids. The findings of German and his colleagues and the Sonnenburgs and their colleagues have shown that many strains of these beneficial bacteria have probably disappeared from the guts of those of us living in the Western world. Keep in mind too that there are many, many types of fiber, and different foods contain different types of fiber. These various types of fiber feed different bacteria that then produce different kinds of short-chain fatty acids. For example, inulin comes from foods like onions, leeks, wheat, and rye; pectin is found in apples, apricots, carrots, and oranges; and fructooligosaccharides are found in bananas, garlic, and asparagus. These fibers feed bacteria that produce acetate, propionate, and butyrate, which represent most of the known short-chain fatty acids in our bodies.[33] But it's not just plant foods; animal foods also feed these bacteria—butter, cheese, and cow's milk help make butyrate in your gut, for example. And the oligosaccharides that German and his team found in breast milk are converted to acetate by *B. infantis*, which appears to be highly protective for babies.

There is still much to learn about the role of the carbohydrates that feed the bacteria in the gut. But out diets of highly processed foods—from birth on—are making us vulnerable to disease in ways we never could have imagined.

Additives in Our Ice Cream

We also must consider the effects of the chemical additives that we put into our bodies as we eat. Some of the most common additives appear to be affecting our microbiota in unexpected ways. Andrew Gewirtz, a researcher and professor at the Center for Inflammation, Immunity and Infection at Georgia State University, is concerned about the emulsifiers found in processed foods—the common additives like carboxymethylcellulose (CMC) and polysorbate 80. Mouse studies by Gewirtz and his colleagues have shown that these compounds, which act like detergents, change the composition of the microbiota and damage the protective lining of the gut wall, leading to chronic inflammatory conditions.

Emulsifiers are used in products to make foods one consistency, or one "phase," as scientists refer to it. Think of natural peanut butter: the oil separates and rises to the top. An emulsifier combines the product into one smooth texture. But what are emulsifiers doing to our microbiota? Gewirtz's studies show that the emulsifiers damage the lining of the gut wall. "The emulsifiers that we have tested are disrupting the composition of the gut microbiota, they're changing the species of bacteria, and they are doing it in a way that promotes inflammation," he said. Gerwirtz and his colleagues have found that the emulsifiers change the microbiota by killing off more fragile species while promoting heartier, more aggressive species. These aggressive species are then eating the mucosal lining of the intestine and penetrating deeper into the tissue, promoting an immune system response that results in chronic inflammation. And the mouse studies have shown that even low doses of these additives can be a cause of metabolic disease and more severe diseases as well.[34]

I asked Gewirtz about the amounts of these emulsifiers and what they are testing versus the amounts in our foods. The Food and Drug Administration (FDA) sets specific rules for amounts of various additives allowed in food products. For the two emulsifiers in question, any given product can contain as much as 2 percent CMC or 1 percent polysorbate 80. "If we give mice a food that contains 0.1 percent polysorbate 80, we are seeing very clear effects," Gewirtz said. Polysorbate 80 is often used in ice cream—to prevent water and fat solids from separating—and the amount that the researchers are testing is equivalent to the intake of a person who eats a pint of ice cream a day. "What the food industry would say is that, 'That is a lot of ice cream to be eating,' but these emulsifiers are in so many products, and we [as researchers] are certainly not exceeding the total exposure that the average person has to emulsifiers in our mouse model," Gewirtz said.[35]

What about the more natural emulsifiers, such as the soy lecithin, xanthan gum, or guar gum often found in many organic processed foods? Gewirtz is looking at those as well and has found that they have effects similar to the synthetic emulsifiers like CMC and polysorbate 80 but to a slightly lesser degree. Does this mean they are safer to eat? Not necessarily. As you'll recall from my Whole Egg Theory, just because an additive is derived from a natural product does not mean that it is safe. (Speaking of eggs, egg yolks are the original emulsifiers, used to make ice cream, salad dressing, and mayonnaise or aioli. Egg yolks are more expensive, harder to work with, and obviously have a shorter shelf life than synthetic, or even "naturally derived," emulsifiers.) For example, guar gum is an additive that appears in an array of packaged foods, including those labeled ORGANIC and NATURAL. It is a chemical that exists in the guar plant, which is commonly eaten in India in its whole plant state. But when the chemical is extracted for use as an emulsifier, the concentrations in which it is being used are dramatically different from the small amounts that exist in the plant. "So even though the product is natural, you'd be really hard pressed to say that the dosage and the context is natural when it is used as an emulsifier. And guar gum is an emulsifier that we are seeing [has] pretty strong effects [on the microbiota] in our models," Gewirtz said.[36]

The issue of combined effects is at play here too. Gewirtz and others have tested emulsifiers, in isolation but not in combination, yet we rarely take in one additive at a time. Many foods contain multiple types of emulsifiers, and if you eat more than one food that came out of a package over the course of a day, you are certainly taking in more than one and in unknown quantities. This last bit about unknown quantities is also important. The FDA sets upper limits for the amount of additives allowed in foods, but we do not know exactly how much of any one additive is in the food. The FDA also does not consider the combined effects of various additives, even in the same food product. What's more, legislation protects manufacturers from having to disclose just how much is in any food item, and they can claim that these details are "proprietary information." Gewirtz, like many of the other researchers I've spoken to, believes this is a crucially important point, and he and his team are planning to look at the effects of common additive combinations in future research.

FOODS WITH EMULSIFIERS

Industrially produced emulsifiers lace our food supply—even organic and natural foods. They are often present in ice cream, soy milk, almond milk, yogurt, peanut butter, ketchup, mayonnaise, salad dressing, processed meats, vegan meat products, processed bread, crackers, cookies, and other snack foods. Some common names for emulsifiers on labels are xanthan gum, carrageenan, polysorbate 80, guar gum, and soy lecithin.

Another common additive that you often find in products that contain emulsifiers are artificial sweeteners. In 2014 a team of researchers in Israel found that the sweeteners, which are promoted as benign and even healthful choices for people who are overweight or diabetic, could actually be causing obesity and diabetes.[37] This is not unlike the trans fat debacle, which turned out to cause the very diseases we were told they would prevent. The Israeli researchers were not the first to note the connection between artificial sweeteners, obesity, and diabetes, but their study is unique because it points toward (but has not definitely proved) causality.

One researcher, Dr. Eran Segal of the Weizmann Institute, told me that the relationship between artificial sweeteners and obesity is well known from epidemiological studies. "People who consume artificial sweeteners tend to have higher weight, and [the sweeteners] are connected with obesity, cardiovascular disease, and metabolic disease; but when you observe that relationship you don't know cause and effects." In other words, do people who are already overweight switch to drinking diet soda to reduce their calorie consumption? Or are they overweight because they are drinking diet sodas? The researchers worked to tease out causality by doing experiments in mice.

The researchers added the three most popular artificial sweeteners, saccharin (the pink packets of Sweet'n Low), sucralose (the yellow packets of Splenda), or aspartame (the blue packets of Equal), to the drinking water of mice and compared them to mice drinking plain water and mice drinking water with regular table sugar. To the researchers' surprise, the mice drinking the artificial sweeteners developed glucose intolerance, which is one of the key markers in diagnosing diabetes. The mice drinking water or regular sugar showed no such problems. Then the researchers gave the mice with glucose intolerance antibiotics. "And lo and behold, after giving antibiotics we cured the mice of their glucose intolerant condition, and because antibiotics mainly act on the microbes, this was a very strong hint that the mechanism through which artificial sweeteners induce glucose intolerance was through an effect on gut bacteria," Segal said.

The researchers then went a step further and performed experiments on "germ-free" mice, which have been bred to have no microbiota of their own. The most compelling of these studies took bacteria from a healthy mouse who never drank artificial sweeteners and placed them in a culture. Then the researchers added an artificial sweetener to that bacteria. They then put those bacteria into a germ-free mouse who had never consumed the sweeteners, and after the transfer of bacteria, these previously healthy mice developed glucose intolerance as well. "This proved that these sweeteners are making some change in the bacteria, and this change alone is inducing the glucose intolerant condition that we observe in the animals when they drink these artificial sweeteners; in other words, they don't need to drink the sweeteners to induce this condition; you just need to give it to the bacteria," Segal said. The researchers also did a small-scale study in humans and for one week gave artificial sweeteners in amounts allowed by the FDA to people who had never previously consumed artificial sweeteners. The researchers found that amount was enough to alter the subjects' gut bacteria and induce glucose intolerance in more than half the participants. The researchers also transferred the human bacteria from those who developed glucose intolerance to the germ-free mice, and those mice developed the condition. "So this was a proof of principle also in humans," Segal said.[38]

A related part of Segal's experiment demonstrated that antibiotics wiped out the bacteria that were causing glucose intolerance in the mice—an example of how antibiotics might be used to treat a dysbiotic gut. In the vast majority of cases, though, the overuse of antibiotics seems to be at the root of an imbalanced gut and is a major part of why so many people in the Western world have a less-than-ideal microbiota. Much has already been written about antibiotics in our food supply, but the two main points to understand are these: feeding farm animals subtherapeutic, or low levels, of antibiotics makes them gain weight faster. This practice is widespread among farmers and is seen as a boon to production and profits. If low levels of antibiotics make farm animals fat, it stands to reason that low-level exposure to antibiotics could be contributing to weight gain in humans as well. Livestock, poultry, and farmed fish are given antibiotics along with their feed, so this means not only are we taking in more prescription antibiotics than ever before, but antibiotic residue in our foods is ubiquitous. Antibiotics are present in milk, meat, cheese, eggs, and fish, as well as in many vegetables, because antibiotic residue in manure used to fertilize crops seeps into the soil and then into the plants we eat.[39]

Furthermore, millions of Americans are exposed to antibiotics in their water supply every day, especially those of us who live near farms or sewage treatment plants. Antibiotics can remain active for months after being excreted by farm animals and have been found still active in rivers miles from factory farms. One study of a river in Colorado found that, of the five sites tested, the only site not contaminated by antibiotics was in a remote area of the mountains. By the time the river had moved through the agricultural urban areas, six of the ten antibiotics the researchers tested for were present. A 2015 study of urban waterways in Milwaukee found multiple strains of antibiotic resistant *E. coli*.[40] This is no small issue: in May 2016 the United States found its first case of bacteria resistant to a "last-resort" antibiotic, which puts us at great risk of what researchers are calling nightmare bacteria that we cannot control with antibiotics. Despite the many dire warnings about antibiotic overuse and resistance from scientists and doctors, in 2015 the FDA reported that domestic sales of antibiotics for farm animals had increased by 22 percent since 2009.[41]

The implications of our long-term, low-dose exposure—which is what you could call the amount of antibiotic residues that are in our food and water—are just beginning to be understood. Martin Blaser, a microbiologist at New York University, found that feeding young mice low doses of antibiotics changed the composition of the microbiota in the mice, as expected. This effect disappeared a couple weeks after the mice received their last dose of antibiotics. But then a curious thing happened: When researchers fed the treated mice a high-fat diet, they started gaining weight rapidly, and this effect was especially pronounced in females. The female mice gained twice as much fat as the mice that did not receive antibiotics but were fed the same high-fat diet. This study highlights the potentially harmful role of giving antibiotics early in life—as is so common now. What does this practice mean for long-term health and weight gain? Blaser, who has written a book on the subject, believes that antibiotics are another missing piece of the puzzle in explaining our skyrocketing rates of obesity and disease.[42] And as we piece this puzzle together, antibiotics, emulsifiers, artificial sweeteners, and our highly processed diets, which are missing vital components, are likely resulting in the elimination of all but the most hearty, aggressive strains of bacteria, which are potentially encroaching on the intestinal mucosal lining, causing inflammation and eventual disease.

I'd be remiss if I didn't also mention the role of pesticide, herbicide, and fungicide residues in our food and how they affect the microbiota. In 2013 researchers in Germany found that glyphosate, the main ingredient in the most commonly used herbicide in the world, Roundup, kills off many beneficial strains of bacteria in the gut, possibly allowing the heartier and potentially pathogenic strains like *E. coli* to proliferate. The researchers were investigating a stark rise in botulism in cattle in Germany and traced it to the explosive increase in glyphosate exposure. Use of the herbicide increased globally from 112.6 million pounds in 1995 to 1.65 billion pounds in 2014.[43] Normal levels of *Clostridium botulinum* are kept in check by a balance of healthy microflora, but glyphosate is also a powerful antibacterial, and the researchers in Germany believe that the decimation of healthy gut bacteria in the cattle led to a bloom of the pathogenic bacteria.[44] Are you noticing a trend here? It appears that many substances that we are taking in with our foods, including Roundup residue on produce and other foods we eat, are having similar effects on our microbiota.

One other key point that Andrew Gewirtz, the researcher studying emulsifiers, noted is that certain elements that are missing from our processed food diets might be as important as what we've added to processed food. This idea is part of my Whole Egg Theory and is crucial to really understanding our current food crises, which operate on multiple fronts. Gewirtz told me about an interesting study that gets at some of the underlying issues and blind spots in animal research.[45] In this study he and his colleagues looked at what mice in laboratory studies typically eat. Researchers often feed lab animals a high-fat diet or a "normal chow diet," which is lower in fat, to demonstrate the effects of diet-induced obesity. What this discussion often omits, however, is that the high-fat diet is also a highly refined, processed food diet—what researchers call a "purified diet." On the other hand, the normal chow diet is composed of a relatively unrefined mix of plant and animal products—in other words, more of a whole foods diet. This means it can be difficult to figure out whether the fat content or the refined versus unrefined nature of the diet is causing weight gain or ill health in the animals.

Not surprisingly, Gewirtz found that it was not so much the high fat content of the feed but rather the degree to which it was processed. The study showed adding more soluble fiber, like inulin, to the animals' diets mostly eliminated the effects of weight gain from the high-fat diet. Adding inulin protected the mice from the loss of intestinal and colonic mass that was prevalent in the mice on the processed high-fat diet. And the beneficial effects of the inulin were a result of a more balanced microbiota that decreased the inflammation that drives obesity and metabolic syndrome. Gewirtz said these highly processed diets promote disease and result in a fragile and weak intestine in the mice. "I think, in addition to being specific food additives that are causing problems, the processing of food is also removing key nutrients that we have yet to identify [but] that are important," he said. "Overall, it will be healthier to eat food closer to its original forms. It's not just the addition of specific chemicals that are a problem, but there are things being lost in the processing—some of those things we have ideas about, like fiber—but there are other ingredients that we don't yet have a handle on." In their study, the researchers could mediate the negative consequences of the processed food diet by adding fiber, but that doesn't get to the root of the problem. "We can correct a lot of these problems in the purified diets by adding a lot of fiber, but that doesn't really tell us what was missing," Gerwitz said. "It may be like patching a tire, but it doesn't tell you what the problem is in the first place."[46]

This idea recalls the findings of both German at UC Davis and the Sonnenburgs at Stanford. The case of the missing *Bifidobacterium infantis* in babies appears to set us up for a variety of autoimmune diseases that trigger inflammatory responses throughout our bodies. German told me that while the science is not yet definitive in humans, it looks like the missing bacteria result in a higher risk for asthma, eczema, and other allergic conditions. Indeed, in much of the research on *B. infantis*, babies with the bacteria present in their stool are healthier and more robust, and in multiple animal studies, the administration of the bacteria prevents the incidence of these various disorders as well.[47] And the bacteria missing throughout our lives are what the Sonnenburgs demonstrated in both their animal models and by observing the Hadza in Tanzania, who eat the foods humans have been eating for millennia, that is, a completely whole foods diet of foraged and hunted foods while enjoying disease-free lives.

We take for granted the "diseases of civilization" such as diabetes, cancer, or autoimmune diseases ranging from autism to allergies to arthritis. Most people in the Western world live with some form or another of these diseases—but it simply doesn't have to be this way. We too could live disease-free lives or at least drastically reduce our risks for them if we eliminated processed foods from our diets from birth and embraced a lifelong diverse diet of whole foods. (At this point we will also have to reintroduce important bacterial strains to our bodies that have gone extinct, such as the *B. infantis* that German and his team discovered. Indeed, they have produced *B. infantis* in the lab and started a company called Evolve Biosystems; the bacteria are now available to all newborns in the United States.) The Hadza population—among the several hunter-gatherer or pastoralist groups still left on the planet—are proof that human beings were not designed to get sick and die of disease at the rates we are seeing in the Western world today.

One argument often invoked when comparing Western health outcomes with those of people living in hunter-gatherer communities or rural agrarian communities is that our life expectancy is longer than theirs. The argument usually goes something like, "Well, we live far longer than those populations so we must be doing something right." But that argument falls flat with just a little bit of scrutiny. In hunter-gatherer societies most mortality occurs within the first five years of life because their sanitation isn't on par with ours, thereby increasing the risk for infections. In addition, they don't have access to antibiotics for true life-threatening infections, or access to vaccinations, so it is understandable that infant mortality rates are high. That doesn't speak to the development of chronic diseases, though, so if you take infant mortality out of the equation and look at populations like the Hadza in Tanzania, you'll see that they are fairly long lived. "And if you look at the reasons why they die, you'll find incidences like someone is in their seventies, climbing up a tree to get some honey, and falls. So, this life expectancy argument is a difficult one to make because of that," Erica Sonnenburg told me. But even if you were to throw all that out for the sake of argument and say that still, hunter-gatherers just don't live as long as we do, we must look carefully at their young populations compared to ours. "I think the argument for why there is something fundamentally wrong with our Western microbiota can be made just by looking at our young population," Sonnenburg said. She pointed to the continued increase in Western diseases in our children and young adults—metabolic syndrome, autoimmune diseases, food allergies, type 2 diabetes, and colon and rectal cancer. "If you look at the young populations of these hunter-gathers, they have none of that; there are no obese children, there is no metabolic syndrome, no allergies—and so the fact is, you have to think there is something fundamentally wrong in the West," Sonnenburg said. "And I think the microbiota and its difference from these traditional populations could be a smoking gun."[48]

PART THREE

New Science and Failing Health Agencies

No picture of our industrial food system is complete without understanding the effects of the approximately nine thousand environmental chemicals that have been added to our foods. Of those, about one thousand have *never been tested* for safety. And for those that have been tested, there are serious issues with testing protocols and standards at the Food and Drug Administration (FDA) and the Environmental Protection Agency (EPA). One key problem at the regulatory agencies is that they do not properly regulate a class of chemicals known as endocrine-disrupting chemicals. These chemicals affect our hormonal systems and are ubiquitous in our food supply—from the plastics in our water bottles to the can linings of our beans. These chemicals appear to be fundamentally changing our bodies. Even beginning in the womb, many of these changes have lifelong consequences for our health and for the later development of disease. As we saw in chapters 3 and 4, what happens in the womb has profound consequences for our health in the future, and the scientific findings about endocrine disruptors are adding a crucially important layer of understanding to this emerging knowledge base. What's more, researchers in the field of epigenetics are finding that exposures to endocrine-disrupting chemicals are changing the way our genes are expressed, and we are passing these changes on to future generations, resulting in a range of maladies—from obesity and diabetes to infertility and cancer. Indeed, some of these findings question the very foundations of scientific knowledge. Yet the FDA and the EPA—the agencies that are tasked with keeping our foods safe—regularly ignore or fail to understand the changing scientific landscape. In doing so, these agencies all too often favor industry over public health protection and are failing to keep us, our families, and our food safe.

6

"So Many Tips of So Many Icebergs"

How Endocrine-Disrupting Chemicals Are Harming Us (and Our Children and Their Children)

My long-lived, disease-free grandmother was most likely not exposed to hormonally active manufactured chemicals—endocrine disruptors—in the womb or as a developing child or teen. But by the time she was an adult and having her own children in the 1950s and later, she certainly was. My parents, both born in the 1950s, definitely were exposed to endocrine disruptors in increasing amounts, and I, as a baby and child of the 1980s and 1990s, was exposed to thousands of relatively new environmental chemicals, from pesticides to food packaging. As a new mother, I wonder what the legacy of three generations of exposure to thousands of environmental chemicals will mean for my baby's health, both immediately and long term. With the relatively new discovery of endocrine-disrupting chemicals, or EDCs, and the science of epigenetics, or the effects over the course of generations, new and alarming findings show how our exposure to thousands of chemicals is harming us—and future generations.

The discovery that many of the chemicals we are consuming every day are EDCs, and are probably changing our bodies, complicates long-held beliefs about genetic determinism—the idea that all our behaviors and traits, including our health, are simply the result of our inheritance, or set in stone in our genes. Even Charles Darwin, who spawned this thinking as the father of evolutionary theory, realized he had made a mistake in his theories about genetic determinism. In 1876 he wrote: "In my opinion, the greatest error which I have committed has been not allowing sufficient weight to the direct action of the environments, i.e., food, climate, etc., independently of natural selection. . . . When I wrote the *Origin*, and for some years afterwards, I could find little good evidence of the direct action of the environment; now there is a large body of evidence."[1]

Yet to this day Darwin's ideas about genetic determinism pervade the medical and scientific fields as well as most people's commonsense knowledge. This has a lot to do with why, as a culture, we've been so slow to acknowledge and adapt to the emerging science about how drastically our changed food environment is changing our bodies—and even altering the expression of our genes. In fact Darwin specifically mentioned food; even in a time before industrial food, Darwin was aware that what we eat matters for us as individuals and for our health as a species.

We now have compelling evidence that we are much more than the sum of our genetics and that our environmental exposures, whether through food, air, water, or soil, have remarkable effects on the status of our health. Some of this is obvious: heredity alone does not account for the skyrocketing rates of obesity or for the steep rise in type 2 diabetes in children, which was previously unheard of. Clearly something else is going on. Indeed, the incidence of nearly all chronic diseases has increased dramatically since the late 1950s, making these diseases more prevalent than infectious disease worldwide.[2]

By acknowledging the importance of environmental exposures, Darwin anticipated the two related fields of science I will explore here: The first is epigenetics, which is the study of changes to an organism as a result of a modification in the way a gene is expressed, rather than changes to the DNA itself. Epigenetics is a broad field and shows that exposures to stress, illness, or trauma can cause temporary or permanent changes to our genes, resulting in changes to both our behavior and personality as well as our susceptibility to disease. And when it comes to what we eat—certain diets, specific foods, as well as too many or too few nutrients—can lead to epigenetic changes. I will focus exclusively on our exposures to environmental chemicals through our industrial food supply because they affect each and every one of us every day. Many of these chemicals have alarming effects on our health, and some are also proving to be heritable—meaning they can be passed on to future generations—for example, a predisposition to obesity and diabetes, certain cancers, and disruptions to fertility and sexual development. As such, epigenetics represents a major paradigm shift in understanding the etiology of disease or health as we recognize the complex interplay between our genes and our environment.

What makes the findings about endocrine disruptors especially concerning is that once exposed, we may carry that health burden for our entire lives, especially for chemicals that are persistent pollutants, meaning they don't break down readily in our environment or our bodies—and we may then pass that health burden on to our children and our grandchildren. One prominent example is dichloro-diphenyl-trichloroethane, or DDT, the first synthetic insecticide, developed in the 1940s. The chemical was widely used for delousing soldiers in World War II, for spraying neighborhoods to control the mosquito population, and for killing pests on agricultural crops. DDT was at first believed to have low toxicity in mammals, but after extensive use from the 1940s to the 1970s, it was eventually found—with the help of Rachel Carson's publication of *Silent Spring* in 1962—to be killing birds and other wildlife and interrupting important interdependent processes like pollination. DDT is still found in the fatty tissues of large sea mammals like seals and is even present in the bodies of Eskimos who are far from any source of contamination. It's also found in human breast milk to this day, even though it was banned in 1972.

DDT, considered a probable human carcinogen, is also a likely endocrine disruptor affecting the reproduction of both animals and humans. After decades of use that have essentially amounted to a large-scale human experiment, we now have clear evidence linking DDT to breast cancer, male infertility, miscarriages, developmental delay, Alzheimer's disease, and damage to the nervous system and liver. Extremely small doses of DDT during pregnancy show major effects on fetal development years after maternal exposures. One of the most alarming findings is that daughters exposed in the womb have *quadruple* the risk of breast cancer after puberty, perhaps pointing to a cause for the steep rise in breast cancer incidence since the mid-1980s in nearly all industrialized nations. According to Breast Cancer Prevention Partners (formerly the Breast Cancer

Fund), between 1973 and 1998 breast cancer rates increased by more than 40 percent.[3] Breasts appear to be particularly vulnerable to EDCs because breast tissue develops into adulthood and changes with each menstrual cycle. The heritable component of breast cancer is only 5 to 27 percent, meaning that at least 70 percent of breast cancers are the result of environmental factors. And because DDT is such a persistent chemical, its reproductive health effects are present in people of child-bearing age today. Even now we are exposed to it in our food supply, mostly in meat, fish, and dairy.[4]

The rise in breast cancer is likely an example of how an epigenetic effect plays out in our bodies. Epigenetic changes profoundly alter how cells and tissues work, often leading to diseases or impaired functions. It works like this: Our DNA in all our cells is identical (except for sperm and egg cells, which have half the DNA); however, the DNA in different types of cells contain instructions for making proteins specific to the regions of the body in which those cells reside. For example, cells in the brain contain DNA that instructs them to make proteins needed to make or repair synapses and neurotransmitters, whereas the DNA in a skin cell tells it to make keratin, the protein that gives skin elasticity. That is, although the DNA in all cells is identical, individual cells have specific directions that enable the cell to carry out its biological functions. One way to visualize these cells is by imagining DNA as string wound tightly around spools, and each spool contains unwound or open regions of DNA, depending on the nature of the cell. The determination of these open and closed regions is under epigenetic control. Because the unwound or open regions are exposed and accessible, natural factors like hormones and nutrition status can activate the DNA and lead to protein production. But this also makes cells vulnerable to environmental exposures, for example, chemicals in our food supply, that can influence how and when genes are expressed. Those changed gene expressions can also be heritable, as in the case of DDT exposure.

Our food supply represents one of the biggest sources of EDC contamination, in the form of pesticides, herbicides, fungicides, and the chemicals in food packaging. As the name implies, endocrine-disrupting chemicals impede the proper functioning of your endocrine system, which is involved in every single aspect of the body, from regulating metabolism to healthy sexual and brain function. The endocrine system works through hormonal signaling; the major glands and organs of the endocrine system include the pituitary and pineal glands in the brain; the thyroid; the pancreas; and the adrenal glands as well as the ovaries and testes. Hormones are chemical messengers, molecules that are made in the glands and released into the blood system; they travel throughout the body, but they target certain cells and specific receptors. These hormones, known as endogenous hormones, work by binding to specific cell receptor sites to create changes in our cells' regulation of everything from brain development and function to mood and metabolism to sexual and reproductive health.

But we are also exposed to exogenous hormones, that is, those coming from outside our bodies—sometimes intentionally, as in the case of birth control pills—or unintentionally through environmental exposures, such as pesticides on our food or plastic food packaging. An endocrine-disrupting chemical is an exogenous substance that interferes with the body's normal hormonal signaling, resulting in developmental, reproductive, neurological, and immune effects in both humans and wildlife. Scientists and researchers have found "strong mechanistic, experimental, animal, and epidemiological evidence for endocrine disruption" and its role in obesity and diabetes, problems with female and male reproduction, hormone-sensitive cancers in women, prostate cancer, and impairment to thyroid function, neurodevelopment, and neuroendocrine systems. The environmental chemicals for which the most evidence exists are bisphenol A (BPA), phthalates, pesticides, persistent organic pollutants such as polybrominated diethyl ethers, and dioxins.[5] These chemicals are the known EDCs—those that have been evaluated for their risk of harm. Of the roughly eighty thousand environmental chemicals in commercial production today, the vast majority are untested for safety despite their ubiquity. How many of these untested chemicals are endocrine disruptors is anyone's guess—but current estimates put the number that meet the criteria for EDCs at about one thousand. The National Institute of Environmental Health Sciences warns that "endocrine disruptors may be found in many everyday products—including plastic bottles, metal food cans, detergents, flame retardants, food, toys, cosmetics, and pesticides."[6]

Low-Dose Exposures

Because endocrine disruptors affect our hormonal systems, it is crucial to understand that our bodies are keenly sensitive to incredibly small amounts of hormones—this is simply how the endocrine system works. A basic tenet of endocrinology is that the endocrine system is exquisitely sensitive to tiny amounts of hormones or hormone-like substances—whether they are hormones we produce inside our bodies (endogenous hormones) or those we are exposed to in drugs or the environment (exogenous hormones). Indeed, pharmacology (the study of how drugs work) and drug development are based on the understanding that all drugs have vastly different dose ranges, and the physiologic effects of each drug do not always depend on the dosage. That is, a tiny amount of a drug can have a powerful effect on our bodies. Think of the birth control pill: it can contain as little as twenty micrograms of estrogen yet prevent a woman from ovulating. (One microgram is one thousand times less than a milligram, the usual measure for vitamins.) Because many of the chemicals we are exposed to through our food supply are at low levels, acute or immediate effects are often not the problem; rather, long-term effects and effects passed down over generations are the main concerns.

Therefore, one of the most important features of EDCs is that they are biologically active in small amounts. This finding completely upends the commonly held wisdom that "the dose makes the poison." In the case of endocrine disruptors, this is simply not true. Yet the way chemicals are tested and regulated allows for small doses in foods and food packaging, or what industry and regulators would call trace amounts. The study of toxic chemicals employed by manufacturers and regulators is called toxicology and much of it is based on the idea that the dose makes the poison, or in other words, that it's not the substance itself but the amount that matters. As a result, regulators have long considered trace amounts in and on our foods safe, or at least thought to not cause harm. But the science of endocrine disruption says otherwise. "You can take a hormone, you can go into the lab and do an experiment to

ANIMAL RESEARCH

Much of the current research on endocrine disruption and environmental chemicals is based on animal studies for the obvious reason that testing various hypotheses on humans is unethical (the ethics of doing so on animals is clearly also problematic). People often debate whether animal findings are relevant to humans. The scientists I've interviewed are comfortable making general conclusions about human health based on the results of animal testing. Bruce Blumberg, a researcher and scientist at UC Irvine, makes the important point that the Food and Drug Administration (FDA) and the Environmental Protection Agency (EPA) base their determinations of the relative safety of food ingredients or chemicals we are exposed to almost entirely on animal models and animal research. Industry often criticizes independent scientific findings by saying they are animal models with questionable relevance to humans while using animal models in their own research to get approvals from the FDA or EPA. "If the industry guys want to say, 'Look, Blumberg, you are doing work in animals and it's not relevant in humans, then I say, 'Well, your safety tests were also done on animals, so you can't have it both ways. Either animal research predicts effects on humans or it doesn't—pick a position and be consistent,'" Blumberg told me in a 2016 interview. Blumberg, along with the many other researchers I interviewed for this book, believes that, generally speaking, effects in animals predict effects in humans. "Unless you are a creationist and believe everything was created de novo and has no relationship, there's a biochemical similarity and effects will be similar," Blumberg said.

R. Thomas Zoeller is a researcher who studies endocrine disruption and the thyroid and has worked with both the FDA and EPA. He says that if the agencies want to say that the findings of independent researchers are not valid because they have used animals as test subjects, why do these agencies' guidelines require the use of animals for safety testing? "Most of the data comes from animal work; it tells us about how the [endocrine] system is working. Most of what we know about the endocrine system, about endocrinology, comes from animal studies," he said. "It is the height of being disingenuous to make the argument that since it is an animal study, it is not relevant to humans."[7]

activate the estrogen receptor in cells, and you'll find there's a level where there is no activity, and then you'll add more hormone and you'll get activity; then you add even more and the activity goes away," said Bruce Blumberg, a professor of developmental and cell biology, pharmaceutical sciences, and biomedical engineering at the University of California, Irvine. The effect he described is what is known as a nonmonotonic, or nonlinear response—small amounts of certain substances can have a major effect on cells and the body, whereas larger doses may have no effect at all. Most of the chemicals we are exposed to every day are in relatively low doses, but because the regulatory agencies do not evaluate them that way, safety decisions are not based on relevant exposures. "It is crucial to remember that studies for the safety of environmental chemicals are done at high doses, not the doses that humans are exposed to," said R. Thomas Zoeller, a biology professor and researcher at the University of Massachusetts, Amherst.[8] It's also important to understand that, for some environmental chemicals, low doses can have an even greater effect on the body than high doses.

BPA, the "Nightmare Molecule" Lurking in Our Beans

One chemical that has a profound effect at low doses is bisphenol A, or BPA, one of the most common and well-researched endocrine disruptors. It is found in much of our food and beverage packaging—including water bottles, water coolers, soda cans and bottles, and food cans used for soups, beans, tomatoes, tuna, and sardines, among hundreds of other foods. In more than one thousand studies in animals and one hundred in humans, the chemical, which is structurally similar to the natural hormone estrogen, shows adverse health effects, ranging from metabolic disruption and increased risk of obesity to infertility, altered brain development, and certain cancers, such as breast and prostate.[9] BPA leaches quite readily into food from plastics and can linings—the Centers for Disease Control and Prevention estimate that 99 percent of Americans have detectable levels of it in our blood.

Frederick vom Saal, Curators' Distinguished Professor and professor emeritus of biological sciences at the University of Missouri, told me that researchers are seeing effects from BPA at extremely low doses. In his own studies he has seen effects at two picograms, that is, parts per trillion (one trillionth of a gram), in animal studies. "That is ten to one hundred times lower than what the U.S. Food and Drug Administration and the National Toxicology program say you would expect to find in humans. And it's a thousand times lower than what is thought to be present in human babies," he said. "We're down to a molecule, a cell that can alter things."[10]

In the late 1990s and early 2000s, advocacy groups demanded that this estrogenic compound be removed from baby bottles and child "sippy cups." Now many baby bottles and other packaging materials contain bisphenol S, or BPS, a chemical with a similar chemical structure—and not surprisingly—related negative health consequences. But manufacturers claim "BPA-Free" on products, giving the illusion that harmful chemicals are no longer present.

Laura Vandenberg, assistant professor of environmental health sciences at the University of Massachusetts and an EDC researcher, said that while there is some indication that BPA levels are going down in the general population, BPS levels are going up. Like DDT, both BPA and BPS are so similar to natural estrogen that they can bind to estrogen receptors in the body. Vandenberg explained it like a key and a lock: estrogen is the key and the receptor is the lock. These estrogenic chemicals fit in the receptors for natural estrogen and then induce biological changes in our bodies. Once the chemicals bind to the receptors, they go to spots on the DNA where it binds again and turns certain genes on or off, creating an epigenetic effect. This is especially harmful to fetuses. Vandenberg has found that female rodents exposed to BPA in the womb exhibit risk factors for breast cancer and as they age are developing full-blown carcinomas in the mammary gland. Vandenberg also sees increased risk factors for cancer in the mammary glands of male rodents exposed in the womb.[11]

During normal development, the fetus produces proteins to protect it from estrogen; these proteins sequester natural estrogen so that it can't get to, or act on, its target. But when a fetus is exposed to certain estrogenic chemicals, like BPA, the proteins are not protective and do not bind to the chemicals, which means these estrogenic compounds make their way to the fetus. "Suddenly you are being exposed to these environmental chemicals that mimic estrogen at a time when you are not normally exposed," Vandenberg said.[12]

BPA is problematic in other ways—not just as an estrogenic chemical. "We've been focusing on BPA as an estrogen, but in addition to being an estrogen, it also turns out to be very promiscuous in interfering with all kinds of biological process," vom Saal said. One biological process that he is most concerned about is the regulation and development of fat cells. Vom Saal and his colleagues found that low-dose fetal exposure in mice led to an increase in body weight, liver weight, and abdominal fat mass when those mice became adult males.[13] And in 2016 another group of researchers found that low-dose exposure to BPA in rodents during the developmental period caused fat accumulation and promoted inflammation resulting in decreased insulin sensitivity in fat cells. "There is nothing about BPA that would have predicted it would be like your ultimate terrorist molecule. This is just a nightmare molecule that should not be in contact with our food," vom Saal said.[14]

The Grandmother Effect

The BPA in your plastic water bottle is affecting not only you but your future children and grandchildren. In dozens of studies the chemical shows transgenerational effects, or effects that have been passed down through the generations. "One of the things that is really scary about chemicals like BPA is that if you are exposed to this when your fetus is developing, you can imprint the egg cells in the female ovary and the sperm cells," vom Saal said. "So you are showing the consequences of things that your great-grandparents may have been exposed to."[15]

This is precisely why I am concerned about the fetus I am carrying. His sexual, metabolic, and brain development are affected not only by the chemicals I come into contact with (I try to avoid them, but often that's not possible) but by what I was exposed to as a baby and child, what his father was exposed to as a baby and child, and what both sets of grandparents and great-grandparents were exposed to. That's a lot of potential routes for disruption. And when a fetus is exposed to certain chemicals at crucial windows of development, effects to the fetus can be irreversible. This also appears to be true at crucial windows of development during infancy and childhood.

Blumberg has also made important discoveries about the generational effects of another common environmental chemical in our foods. He coined the term *obesogen* in 2006 for substances that cause weight gain regardless of caloric intake.[16] His work has shown that exposure to tributyltin, or TBT, a known endocrine disruptor, results in weight gain through noncaloric pathways—meaning that the animals in his studies gain weight independent of calorie consumption. TBT is an organotin, a diverse group of persistent organic pollutants found throughout the environment. It was first used in the 1960s in paints and coatings for boats to prevent organisms from attaching to the surface (these substances are known as antifouling agents); this use has since been banned, but it was used globally on a massive scale for decades. As a result of this practice, the chemical is pervasive in much of the fish and shellfish we eat today since it leeched from the paint into our oceans and other waterways. The concentration of TBT increases as you go up the food chain in a process called biomagnification. As a result, the chemical has been found to negatively affect the health of sea otters, dolphins, and whales, among many other sea creatures. TBT and other organotins are also used in the linings and sealings of food cans, in polyvinylchloride (PVC) plastics, as fungicides and pesticides on crops, as slimicides in industrial water systems, and as wood preservatives. Like many other classes of chemicals, organotins were wrongly deemed environmentally safe for many years—and they appear to be everywhere in our environment.

Blumberg said that organotins change how the body responds to calories. "The ones we study . . . actually cause exposed animals to have more and bigger fat cells. The animals that we treat with these chemicals don't eat a different diet than the ones who don't get fat. They eat the same diet. . . . They're eating normal food, and they're getting fatter," Blumberg said.[17] Blumberg has also shown that the TBT-induced weight gain is heritable, affecting as many as four future generations. In a 2013 study Blumberg and his team exposed pregnant mice to TBT and found the next three generations developed greater fat deposits, larger and greater numbers of fat cells and fatty livers, and they showed changes to the liver consistent with nonalcoholic fatty-liver disease. Keep in mind that only the first generation was directly exposed to the chemical in utero; the second generation was potentially exposed to the chemical through germ cells of that first generation (germ cell refers to a cell that eventually becomes part of the fetus). But in the third generation the mice showed these health effects without ever being exposed to TBT at any time.[18]

Rates of obesity have certainly shot up in recent decades, and notably, there has been a stark rise in nonalcoholic fatty liver disease, especially in children, where it was once seen only very rarely. My sister, Dr. Sarah Wartman, a surgeon who sees many livers in the operating room every day, has told me that the increase in fatty-liver disease is shocking. "Pretty much without fail, whenever we go in to remove a gallbladder in a patient, we see a fatty liver," she said, adding that it is actually remarkable now to see a healthy liver in her patients. The statistics bear this out: the rate of nonalcoholic fatty-liver disease is now higher than the rate of the disease in alcoholics. It is clearly related to obesity and diabetes, with 70 percent of cases occurring in patients with diabetes; however, there is also evidence that it occurs in individuals who are not obese. Blumberg said that although his studies have yet to prove his suspicion, he believes TBT is causing fatty-liver disease directly, regardless of weight gain or diabetes.[19]

To clearly understand the transgenerational effects of TBT, recall the concept of epigenetics and the image of the spool of string: it is not the DNA sequence that is being changed, but rather environmental chemicals are changing the way the genes are expressed. "Which genes are being stimulated to be expressed or activated? Which genes are being inhibited? These fundamental properties can change in the fetal germ cells," Andrea Gore, a professor of pharmacology and toxicology at the University of Texas, told me. "This is what we think is being transmitted to future generations."[20] It could be happening this way: Your maternal grandmother was exposed to TBT while your mother was in the womb. While your mother was developing as a fetus, you were developing as germ cells within her ovary. This means that not only was your mother exposed to the chemical, but you were exposed to it as a germ cell. What's even more alarming is that your future children (the fourth generation) are also affected, even though your children had no direct exposure to the chemical. The first findings of transgenerational effects came from the scientist Michael Skinner and his colleagues in 2005. After an experiment found developmental effects in the offspring of rats treated with the agricultural fungicide vinclozolin, an accidental breeding among the test animals produced a second generation of rats. The researchers then figured they might as well breed these rats to a third generation and analyze them as well. What they found in this third generation set the stage for our current understanding of the "grandmother effect."[21]

Three generations after the pregnant rats were exposed to the fungicide, the rats had abnormally low sperm counts, which was confirmed in fifteen subsequent studies. The researchers then went on to test more substances in transgenerational rat studies and found problems with the prostate, kidneys, ovaries, and immune system. "[Skinner] really found a lot of serious problems in these offspring," Gore said. Gore collaborated with David Crews, Skinner, and others for a 2007 study that looked at mate choice in third generational animals descended from grandmothers treated with the fungicide compared to rats that were untreated. Untreated female rats had a strong preference for male rats whose ancestors were also not treated with vinclozolin. Gore said, "It is almost certain that there is an epigenetic transmission that makes those males less attractive."[22]

These epigenetic changes speak to the types of changes that Darwin hinted at when he wrote of his developing awareness that it is not just the gradual process of natural selection but also environmental factors causing changes to organisms—which represents a completely different understanding of human health and our relationship to the environment. Vom Saal explained it like this: "Rapid changes in species have been known to occur, but no one using the gradualist approach of natural selection had a rational explanation for it. When you take chemicals like BPA that can hit major developmental regulatory systems and alter them, you create an entirely different phenotype, or set of traits, and that becomes fixed, it doesn't dilute out," he said.[23]

Sexual Development and Building the Brain

All the EDC researchers and scientists I spoke to emphasized the delicate nature of our hormonal systems, which have been primed to respond to exquisitely low doses of hormones, especially at crucial stages in our development. Some of the most alarming findings about EDCs are those that show permanent changes to the sexual development of the fetus—both the development of the reproductive system and the development of the brain, which are, of course, interconnected.

First, it is helpful to understand the process of sexual development in the fetus. Early in the development of mammals our gonads are undifferentiated—they are neither male nor female. Depending on your genetics, with few exceptions you have either XX or XY chromosomes, and the gonads develop into ovaries or testes, respectively. This is known as sex determination, which is entirely governed by genes. However, once the gonads begin to become either ovaries or testes, at around six weeks of gestation, hormones begin to play a role in what scientists call sex differentiation. "This is where EDCs become really important and relevant," Gore said.[24] She studies the effects of environmental chemicals on the developing fetus, in particular effects on the brain, because so much of sexual and behavioral development is dependent upon hormone action in the brain.

Testicular hormones are crucially important to the normal development of the male reproductive tract and genitals—both the internal and external structures that are necessary for reproduction. Notably, the entire body, including the brain, is simultaneously exposed to testicular hormones, and this sculpts how the brain develops. In fact, exposure to testicular hormones is a key factor responsible for masculine brain development, along with the hormones' more obvious effects on the male reproductive system.

In the female, the situation is a little different: the ovaries also start to produce hormones but not at the same high levels relative to the male. These hormone levels are what scientists call quiescent. This means that the female's exposure to hormones is relatively low, and the female reproductive system—the uterus, fallopian tubes, cervix, and vagina—develop in large part under the influence of low levels of ovarian hormones. The normally low levels of ovarian hormones, and the absence of testicular hormones, trigger specific developmental pathways that result in the development of a healthy female reproductive tract and genitals and cause the brain to be feminized. However, if you expose a female fetus to male levels of hormones, as researchers have done in animal experiments, the female becomes masculinized, Gore said. In animal studies a female with a masculinized brain will behave more like a male, mounting other females in a reproductive encounter or rebuffing the efforts of a male to reproduce.[25]

Exposing animals to EDCs disrupts both male and female sexual development. In males, this could mean testicular dysgenesis syndrome, which can manifest as undescended testicles, abnormalities of the penis, poor semen quality later in life, and testicular cancer, among other problems. Since the 1990s scientists have been wondering whether humanity is nearing a serious fertility crisis since male infertility is increasing globally. Various studies show that EDCs are affecting the quantity and quality of sperm—from their ability to swim well enough to reach an egg to affecting the health and life span of the children they create. Indeed, researchers reported in 2017 that sperm count in Western men had *dropped 59 percent* since 1973.[26] And if these trends continue, by the year 2060 the majority of men in the United States and Europe will be infertile. In females, researchers are finding ovulatory cycle disorders, changes to the timing of puberty, and earlier menopause in women. "There is some sort of disruption during the time when the reproductive tract is really sensitive; and then when we become reproductively mature, all of a sudden we discover problems like infertility or early menopause," Gore said.[27]

Gore has looked extensively at a class of industrial chemicals known as polychlorinated biphenyls (PCBs). These chemicals are another example of persistent organic pollutants, like DDT. Although they were banned in 1979, they are still found in the umbilical cord blood of newborn infants today. Industry used them for decades as lubricants, in transformers, and in dozens of other industrial applications, and now these chemicals contaminate the globe. "Even today PCBs are out there, and that's because many of them are persistent chemicals that were designed that way for their uses in industry. It wasn't recognized when they were first used that they would be causing biological effects," Gore said. According to the World Health Organization, more than 90 percent of human exposure to PCBs comes from our food supply: they are present in animal products, from meat and milk to fish and shellfish, and are found in nearly every living creature on Earth.[28] PCBs accumulate in the tissues of animals as they go further up the food chain, another example of biomagnification. Humans, of course, are at the top of this food chain. PCBs, like many other endocrine disruptors, are fat loving, which means they accumulate in the fatty tissues of our bodies, often camping out there indefinitely.

This group of chemicals is one of the most toxic compounds we are regularly exposed to; they cause reproductive and developmental problems, damaging the immune system, interfering with hormones, and causing cancer.[29] In fact, the biotech company Monsanto paid the largest known settlement in U.S. history—$700 million—to plaintiffs in a class action suit back in 2003 for their exposure to PCBs in their community of Anniston, Alabama. Monsanto's facilities were in the heart of the city's black community, where they produced 680 million pounds of PCBs over four decades. A minister in the community said, "I've buried at least a hundred people who died of cancer, many young people between twenty and forty."[30]

Cancer is not the only concern with chemicals like PCBs. When fetuses, babies, and children are exposed to even small amounts of chemicals like PCBs, the health effects are profound. Gore and her colleagues have found that exposing pregnant rats to a low-dose mix of three types of PCBs (there are 202 types of PCBs in our environment) at a time equivalent to midpregnancy in humans causes changes to reproductive and social behavior in the offspring.[31] PCBs interfere with the signaling of both estrogen and androgen during a crucial fetal development period. "The animals look, and for the most part behave, like normal animals, but when you run them through more complex behavioral tests, we see changes in reproductive and other types of behaviors," Gore said. In one study Gore and her team found that female rats exposed to PCBs in the womb either took longer to mate or did not mate at all.[32] (Of course humans display a wide range of behaviors, and the idea of "normal behavior" is certainly problematic. When scientists talk about normal behavior in animals, however, it is without the societal and cultural expectations regarding human behavior and is instead meant to address reproductive behavior.)

Gore and her colleagues also found that PCB-exposed animals exhibit anxiety-like behavior, represented by changes in interacting with other animals, which is obviously part of social and reproductive behavior. In one study Gore used a light-dark box. Researchers placed the rats in a two-chambered enclosure, with the chambers separated by an opening that allows the rat to move freely between them. One side is darkly lit, the other side brightly lit. Normally, rats prefer the dark chamber, presumably because they are better camouflaged and feel safer, Gore said. "There is a natural sex difference in this behavior, with females typically spending more time in the light and being more active than males . . . but this sex difference vanished in the prenatally treated PCB rats," Gore said. "Males given PCBs showed behaviors that were more similar to the females than to the control males." Gore said that the levels of PCBs that rats are exposed to in her studies are comparable to what is detected in humans. Beyond the dose, "exposures in our rats are much more limited than what humans are exposed to, because humans are exposed to PCBs throughout their entire lives, whereas my rats only receive two brief exposures during embryonic development."

Puberty is another sensitive period for sexual and neurobiological development—and exposures to EDCs at this time can cause further disturbances that also are affected by what preceded them earlier in life, especially during prenatal development. As we enter puberty, hormones surge and begin to act on the same tissues and cells that were organized early in life. "Those tissues have been sitting there patiently waiting to be exposed to our natural gonadal hormones so they can become active and make you into an adult," Gore said. Scientists refer to the pubertal period as the activational period, when normal hormones are released in increasing amounts and activate those target cells established early in life. "This activation includes all of those processes that we need to be able to function as adults, to be able to reproduce, to have the maturity to be able to take care of offspring and, ultimately, the survival of the species," Gore said.[33]

During the fetal development of the brain, hormones organize the brain, and the sexes end up with structural brain differences. And ideally at birth, babies have an organized brain structure that correlates with their sex. But if the developing brain was not organized in this way because of a disruption caused by an environmental chemical, the behavioral and functional differences usually are not apparent until adolescence and puberty. Normal hormone exposure is so crucial for the brain's development in a typical male or female manner that when a mutation to the androgen receptor occurs in a genetically male fetus (with XY chromosomes), he becomes insensitive to testosterone, and this genetic male is born looking exactly like a female; the reproductive tract and gonads never became masculinized, and these individuals therefore identify and behave as females. Thus exposures to hormonally active substances through our foods and food packaging at crucial stages of human development appear to be causing changes to sexual development and behavior.

To evaluate the human reality of multiple windows of exposure, Margaret Bell, in another study conducted in the Gore lab, exposed rats to PCBs both prenatally and when they were juveniles, or at about the time that humans would be going through puberty.[34] They found "pretty big differences in sex-specific behavior," Gore said. For example, rats communicate their emotional status, such as anxiety, pleasure, or discomfort, through vocalizations, and the nature of these calls was disturbed in PCB-exposed females. What's more, the second exposure to PCBs changed the effects of the first exposure in complex ways.[35] Gore said this shows that exposure to various EDCs may manifest differently depending upon the age of the subject. "Brain 'reprogramming' caused by developmental exposures to PCBs induces changes to pathways that undergo dynamic changes throughout life—including expression of genes, manifestations of behaviors, physiological effects, etc. If you test or examine the animals at one age, you'll find a suite of changes that may not be the same as the types of changes seen at another age," Gore said.[36] This is important because EDC exposure in humans is a continuum, as Gore put it, that clearly involves many more chemicals and many more windows of exposure than researchers are testing for.

Bruce Blumberg also discussed organizational changes in the brain and EDC exposure when we spoke. "If you were to have a child, what would your blood levels of TBT be when you were pregnant? We don't know because it is not monitored; but the period when you're carrying that fetus is a very, very sensitive time that can be easily disrupted," he said. "And if you disrupt function, then the disruptions are permanent." R. Thomas Zoeller, who studies thyroid and endocrine disruption, likens brain development in utero to building a house, wherein if you lay the concrete floor down before putting in the plumbing, you can't then go back and redo it.[37]

This brings up the important point of irreversibility—once an environmental chemical has altered brain development or sexual development, the changes are permanent. "If thyroid hormone action is disturbed during development, you can't even think about reversibility," Zoeller said. The thyroid is critical for brain development and for control of your metabolism. Thyroid hormones are also essential for proper development in the womb and at birth—every pregnant woman and every newborn is evaluated for thyroid hormone level because it is so crucial for healthy brain development. Zoeller has found that disruption to the thyroid by EDCs is likely resulting in an array of effects on brain development, including impaired cognitive function, lowered IQ, and visual and perceptual disturbances. He also sees issues with attention disorders and even elements of autism spectrum disorder. What's more, traditional testing by doctors does not pick up changes made to the thyroid by EDC exposure. That means a thyroid test could show normal hormone levels, leaving significant impairment to thyroid function undetected. "So here we've got this incredibly important system, and there are chemicals that seem to be able to interfere with it in stealth kind of fashion," Zoeller said.[38]

Women Are Uniquely Sensitive

Another part of sexual development that Andrea Gore has looked at is length of the reproductive life cycle, and she believes that our exposure to certain EDCs is potentially shortening the amount of time that women are fertile. "There is increasing evidence that there are correlations between higher amounts of environmental chemicals detected in women and earlier menopause, so it looks like the whole reproductive life span is getting shortened at its end," Gore said. Gore and other researchers believe that EDCs are also associated with early puberty. Many EDCs stimulate estrogen pathways in the body, which could explain why we see early puberty in girls but not boys, who don't rely on estrogen for puberty. "So we might be looking at an earlier puberty but also an earlier menopause," Gore said.[39] Some studies show menopause occurring a year or two earlier—that may not sound like a lot, but if you are a woman who has postponed having children, early menopause can be devastating.

In one study, Gore and her colleagues looked at a pesticide called methoxychlor that is commonly used on food crops; exposure to this pesticide in the womb resulted in animals that had premature reproductive aging and changes in the brain related to the control of reproduction. These changes were still present in animals when they reached old age—the equivalent of a seventy-five-year-old woman. "What that shows is how permanent these early life changes are, that you can detect them in the brain of aging animals and presumably humans," Gore said. Methoxychlor is not the only substance that has shown a shortening of the reproductive life span. Others, like the BPA found in all our food packaging and the dioxins that come from fuel and waste emissions, have shown similar effects. Studies in human populations have also shown that exposure to diethylstilbestrol (DES), a synthetic estrogen used as a growth stimulant in many food animals, as well as perfluorocarbons (PFCs), which are potent greenhouse gases, during development are linked to earlier menopause.[40]

From the 1940s to the 1970s, an estimated five to ten million pregnant women and their fetuses were exposed to DES in another unintentional multigenerational experiment.[41] At the time, this powerful synthetic estrogen was believed to prevent miscarriage and other pregnancy complications. We now know that exposing the developing fetus to this endocrine-disrupting chemical had tragic health consequences. Many girls exposed in the womb, called "DES daughters," had abnormal development of their reproductive tracts that later led to problems like premature birth, ectopic pregnancies, and miscarriage; others went on to develop a rare form of vaginal cancer.[42] Sons exposed in the womb were shown to have testicular abnormalities, including undescended testicles and increased risk of cysts, infection, and inflammation.[43] Preliminary findings of a study conducted by the National Cancer Institute of the grandchildren of DES mothers show that this generation may have increased risk of cancer, infertility, and birth defects.[44]

Laura Vandenberg has studied the link between EDCs and endometriosis. Endometriosis is the third leading cause of infertility among women, and it is also linked to an increased risk of ovarian cancer and heart disease. As any woman with endometriosis can attest, it can also cause debilitating pain, greatly affecting quality of life and leading to depression and anxiety. "Endometriosis is not killing you—you're just miserable, or you can't have babies," Vandenberg said.[45]

You may have noticed even after this brief survey of the science of endocrine disruption that women, fetuses, babies, and children seem to be particularly vulnerable to the damaging effects of EDCs. I asked Vandenberg, who looks at women's health issues specifically, if she thinks that women are more vulnerable to endocrine disruptors. "In the womb males and females are equally sensitive . . . but adult women versus adult men are more sensitive. Women have [menstrual] cycles, which does mean that we are always in a state of flux, and I think it does lend toward increased susceptibility," she said. "I also think that the fact that we are the ones that carry the next generation means that our sensitivity is not just about us, it is about the next generation."[46]

We must also always remember that the vast majority of the

chemicals to which we are exposed through food or environment have never been tested. In fact, we don't even know what most of them are. "Some of them may or may not be EDCs, but most of them haven't been tested, so the tip of the iceberg in this conversation is that there are actually so many tips of so many icebergs because we don't even know what chemicals are out there," Gore said. "We're looking for what we know. We can't look for what we don't know."[47]

Thanks to the tireless work of academic scientists like Blumberg, Gore, Vandenberg, vom Saal, Zoeller, and their colleagues, what we now know about the ubiquity of certain environmental chemicals and their deleterious effects on our health is quite clear. Also, thanks to their work we have information about epigenetic changes, another paradigm shift in our understanding of human health and the health of future generations. And when we combine the findings about EDCs with their potential to make epigenetic changes, it is likely that we are only just beginning to see the kinds of profound effects recent generations' exposures to so many newly made environmental chemicals are having on us as a species. Could our exposure to the BPA in a water bottle or can of beans, or the TBT in our seafood and the linings of canned goods, or the PCBs in our milk and meat be altering the expression of our genes, and our children's and grandchildren's as well? Could this set us and them up for obesity, infertility, altered sexual development, and certain cancers? Could your mother's or grandmother's exposure to DDT be affecting you now?

The scientists I spoke to expressed great frustration at the current state of our regulatory processes and with the many people in science and medicine who are slow to recognize the changing science. Echoing Darwin's thinking, Blumberg said, "Only about 15 to 18 percent of human diseases have currently identified genetic causes, despite tens of billions of dollars spent chasing them. I predict that environment (broadly defined) acting on the epigenome will turn out to be at least as important as genetics in human disease etiology."[48] Gore agreed. "I never believed that genes were everything because, if you look at any complex disease, most of those diseases don't come down to a single gene," she said. "Epigenetics is everything because we are the combination of our genes and our environment."[49] Our genes may set us up for various predispositions or tendencies, but the field of epigenetics shows that our environment is equally important—if genes are the gun, the environment pulls the trigger. Do we want to live in an environment riddled with toxic environmental chemicals that harm not only us but our children, grandchildren, and great-grandchildren?

The truth is that we do live in that environment, and even with hindsight we haven't heeded the findings of the tragic health consequences of chemicals like DDT and PCBs, which were assumed to be safe at the time. Instead, after banning those chemicals, we replaced them with other chemicals, many of which also are presumed safe, some for decades. The most prominent example that we are all exposed to on a daily basis—what some call the DDT of our time—is glyphosate, or the main ingredient in the weed killer formulation Roundup. Long considered a safe alternative to the more toxic pesticides and herbicides, it laces our food supply. In 2015 the World Health Organization classified glyphosate as a probable human carcinogen. It is the most commonly used herbicide in the world, first introduced in 1974 by Monsanto, and its use has increased one hundredfold since the late 1970s. Americans have applied 1.8 million tons of the stuff since its introduction. Worldwide, more than nine million tons have been sprayed. Glyphosate is applied to nearly all the wheat, corn, barley, canola, oats, and soy grown in this country—and in one form or another, these make up the foundation of our entire food supply. There is also concern that glyphosate is not only a carcinogen but an endocrine disruptor, which means that even trace amounts on our foods, which the regulatory agencies allow, could result in significant negative health effects. Current research points to the ability of glyphosate-based herbicides to disrupt hormonal systems and gene expression at various doses, including low doses.[50]

We do not have the data that tell us how much is on our foods because the FDA has never monitored or tested for it—even though we've been regularly exposed to the chemical since 1974 and our exposure keeps increasing. In February 2016, the FDA announced that it would begin testing crops for glyphosate residue, but by November it had suspended testing with no indication of when (if ever) it would resume. Not only is glyphosate used on genetically modified crops like corn, soy, cotton, alfalfa, canola, and sugar-beets—known as Roundup-ready since they are engineered to withstand the herbicide—it is also used on non-GMO crops, like wheat, barley, and oats (among other crops) to dry them out and hasten the harvest. That means that any nonorganic bread you buy is probably made with wheat treated with glyphosate. Even almonds, pecans, walnuts, oranges, avocadoes, grapes, and apples may have been treated with the herbicide. Roundup is present in animal feed too; it is sprayed on corn silage, alfalfa hay, and sprouts, which are fed to animals in confinement. Therefore trace levels of glyphosate can also end up in your body through the meat and dairy you eat. Perhaps recognizing the multiple routes of exposure to the herbicide, Monsanto recently requested that the allowable levels of glyphosate residue be increased for many foods, including animal feed.[51] (Are you noticing a trend here? Monsanto made DDT and PCBs and is the maker of Roundup.)

How is it that we are living in a veritable soup of environmental chemicals, a great many of which are coming to us in and on our foods, and many of which are untested and could be causing us serious harm? And why is there such resistance to adapt to our changing scientific knowledge that challenges long-held beliefs like the dose makes the poison or genetic determinism? Ultimately, this is because the regulatory agencies that were put into place to protect us are deeply conflicted. The agencies are not living up to their prescribed duties to protect the public health by evolving with the latest scientific findings, keeping industry in check, or keeping us and our children safe from the dangerous environmental chemicals we are all exposed to every day.

Our "Safe" Exposure to Toxic Chemicals

How the Regulatory Agencies Are Failing to Protect Our Health

There was a big news story in July 2017 in the Well section of *The New York Times* with a headline that read: "The Chemicals in Your Mac and Cheese." Researchers found plasticizers, known as phthalates, in the popular kids food. Then, less than two weeks later, the *Times*'s Food section reported that traces of the herbicide glyphosate, the main ingredient in Roundup, had been found in Ben & Jerry's ice cream.[1] Several people asked me about it: Should we be worried? Yes, we should be worried, but not just because researchers found plasticizers in our mac and cheese or herbicide in our ice cream. We should be worried because these kinds of environmental chemical contaminants are literally everywhere, lurking in nearly all our foods. We know they exist in these two foods because researchers specifically looked for them. Roughly nine thousand environmental chemicals on the market end up in our foods, including food additives, colorings, flavorings, pesticides, and food-packaging chemicals.[2] They are ever present in our environment and in our bodies, and the regulatory agencies whose job is to test and monitor them are failing to protect us. In dozens of interviews with scientists, researchers, and professors, I have identified major roadblocks to the fair and impartial regulation of environmental chemicals that we are all in direct contact with on a day-to-day basis, especially those in our foods.

Two main agencies are responsible for regulating all the environmental chemicals in our foods: the Food and Drug Administration (FDA) oversees all food and food ingredients on the market (except meat, poultry, and certain processed egg products, which are overseen by the U.S. Department of Agriculture), and the Environmental Protection Agency (EPA) regulates pesticides, herbicides, insecticides, and fungicides used in food production. But the laws governing the agencies have myriad problems, as do the agencies themselves. As a result, we are not being protected from the essentially unregulated industries that manufacture all the substances that end up in our food and in our bodies.

Chemical Manufacturers Provide Their Own Data

One of the fundamental flaws in the design of the regulatory system is that the manufacturers of the pesticides, emulsifiers, and plasticizers that are added to our food supply, food products, and food packaging provide their own data conducted by scientists that they hire to perform safety testing on their products. Let me repeat that: The manufacturers of the substances that lace our food supply and make their way into our bodies provide the safety data and their own evaluations to the regulatory agencies for review. In the case of pesticides, an EPA spokesperson told me that this practice is intended to save taxpayers money on expensive research and scientific studies. "Congress placed this obligation on the pesticide registrant rather than requiring taxpayers [to] fund data development," the representative said. Most people would assume that the governmental agencies are doing independent testing to ensure the safety of our foods and other products in contact with our foods. But that is simply not the case. Although the EPA and FDA sometimes do review the findings, no independent body or group of scientists tries to replicate the findings. "The reality is that this legislation has no business being there; the only entity testing the safety of the components is the company—and they have no incentive to show that their products are dangerous," said Bruce Blumberg, a professor at the University of California, Irvine, who researches endocrine-disrupting chemicals. "It is a conflict of interest, and they are not protecting public health—that is the bottom line."[3]

Toxicology and "the Dose Makes the Poison"

When we eat that mac and cheese, that Ben & Jerry's ice cream, and probably hundreds of other foods, the FDA and EPA say that since we are only being exposed to "trace" amounts of phthalates or glyphosate, for example, these chemicals are harmless. This strikes many people as common sense—people often say, "If I'm only getting a little bit, it can't hurt, right?" But this thinking, by both the regulators and eaters, is fundamentally flawed based on the groundbreaking discoveries in the study of endocrine-disrupting chemicals, or EDCs. As I discussed in chapter 6, endocrine-disrupting chemicals act on the endocrine system, which is highly sensitive to tiny doses of hormonally active substances, which is what many of these chemicals are. Yet the manufacturers base most of their testing on toxicology, a field that faithfully abides by the outdated edict that the dose makes the poison. Both agencies have been incredibly slow to adapt to the scientific findings that show otherwise. This means that most of the substances in our foods are not tested for their low-dose effects on our bodies, so we have no way of knowing how they are affecting us. "They do not test chemicals at the concentrations that people are being exposed to—period," said R. Thomas Zoeller, a biology professor and researcher at the University of Massachusetts, Amherst.[4]

Of course, some scientists are studying these low-dose effects, but neither the FDA nor the EPA is considering their findings. And although a representative of the EPA told me that they do in fact "consider low doses," many leading experts in the field say the agency does not. "They [the FDA and EPA] may say they consider them, which means they thought about it. Do they ever test low doses? The answer is no, they don't; they start with high doses and then they come down, down, down until they find no effect," Blumberg said. The problem is that for many substances, dose and toxicity are not linear and they can exert strong biological effects at very low doses. Blumberg is referring to the type of testing now conducted by the manufacturers of chemicals, which is a test of gross toxicity, or testing for acute poisoning and death. The manufacturers' researchers start with the amount that causes death in lab animals and then, depending on the health outcome or system of the body they are going to evaluate, decrease the dose until no effect is observed. On the other hand, researchers like Blumberg and his colleagues, who study endocrine disruption, begin with zero and then add tiny amounts of the test chemicals until they see effects. Blumberg said that the levels of chemicals he works with are often lower doses than what industry uses by a factor of 1,000.

And as long as toxicologists continue to test at high doses and operate under the notion that the dose makes the poison, not much will change at the regulatory agencies and we will fail to be protected from the low-dose exposures we are exposed to—whether it's the trace amounts of pesticides on apples or bisphenol A (BPA) in a water bottle.

Maricel Maffini, a scientist and independent consultant who has authored several prominent papers about the problems at the FDA, says that the agency's internal problems stem from a generational divide among its scientists. "Many of their scientists are old school, many of their scientists have told us to our faces that endocrine disruptors are not a thing, that low-dose effects are not a thing . . . they forgot—even though some of them are pharmacologists, and so they are very familiar with the concept—[that] you have to find the sweet spot for the drug to work. But they don't apply the same science when it comes to food additives," she said.[5]

Blumberg makes the related point that the mind-set of toxicologists is to blame. He described it as a kind of denial on the part of the toxicologists and the agencies that depend on the toxicologists' work. "The toxicologists don't want to know about the science of endocrine disruption because it conflicts with their worldview," he said. "Their worldview is that the organism is a black box and you throw things at it, and either bad things happen or they don't—and that's all they want to know for the most part."[6]

This thinking is similar to the scientific notion of energy balance, or "calories in, calories out." For decades, the scientific and medical establishment told us that cutting calories and increasing activity levels would result in weight loss, but new research shows that it is not that simple. The body metabolizes different foods differently even if they have the same caloric content. Much of this is common sense: eating five hundred calories of potato chips will not have the same effects on your body as eating five hundred calories of almonds. But scientists, researchers, and even journalists and authors who have based their careers on the idea of energy balance stubbornly stick to "calories in, calories out," even when presented with new scientific findings that contest this seemingly iron-clad fact. And because endocrine disruptors have been shown to alter metabolism and promote weight gain regardless of caloric intake, both "the dose makes the poison" and "calories in, calories out" are related and outdated modes of thought. Both lines of reasoning work well for the food and chemical industries though because they absolve the manufacturers of responsibility and place the burden on the consumer—demanding that we make healthy choices about what to eat or avoid. But the logic doesn't hold.

Both the EPA and the FDA claim they evaluate chemicals at a

range of doses, including low doses, but how seriously? Maffini told me that when the manufacturers' contract laboratories test chemicals at high, medium, and low doses and find an effect at the middle dose, the manufacturer can say that the result must be unrelated to the chemical—and because the manufacturer is reviewing the findings, they can put them in the best light possible. "I've seen many companies just dismiss the data," Maffini said. One example is an additive called BioPerine, a type of patented black pepper extract that sounds relatively harmless. But its manufacturer, which interpreted previously published data and presented it to the FDA, concluded it showed "no dose-related effect" on organs in animal studies. Yet the actual data show there were indeed effects seen in organs at different doses. Among other more complex findings at all doses, the effects at the highest doses were significant decreases in the weight of the spleen, thymus, and lymph nodes.[7] "The data were not exactly well interpreted," Maffini said.

All the researchers I interviewed agreed that current evaluations conducted by regulatory agencies are faulty for another reason: the agencies are not considering sensitive end points, or systems of the body that are affected by hormones, such as the thyroid, reproductive, or neuroendocrine systems. Nor are they considering fetal development in the womb, especially brain and sexual development. For example, when it comes to environmental chemical effects on the thyroid—which is crucial for brain development—the agencies are only measuring two endpoints: hormone levels in the blood and what's known as histopathology, which includes examining the tissue of the thyroid and measuring the weight of the thyroid. However, examining the gland itself is not an accurate way to measure thyroid hormone action, according to Zoeller, a researcher who studies endocrine disruption and the thyroid. The thyroid system is not only the gland and the hormones it produces but the effect of these hormones (or environmental chemicals mimicking these hormones) on the rest of body. This means that many chemicals in our food supply appear to be affecting the thyroid system without affecting the thyroid gland itself and are slipping through the cracks of the regulatory processes.[8]

Zoeller has been studying the effects of endocrine-disrupting chemicals on the thyroid for decades. He has looked at polychlorinated biphenyls (PCBs) and dioxins that are prevalent in our food supply and shown that because these chemicals are structurally similar to thyroid hormones, they are having profound effects on the thyroid—especially in the womb. The effects include lowered IQ, impaired brain development, issues linked to attention deficit/hyperactivity disorder and autism, and visual and perceptual disturbances. Recall that Andrea Gore's research is showing that PCBs are significantly altering both brain and sexual development in animal models in the womb, and Bruce Blumberg has found epigenetic changes that affect as many as four generations after initial exposure in the womb to tributyltin, or TBT, which is also prevalent in our food supply. Remember: these are only the chemicals that researchers are testing for, but thousands more exist in our food and food packaging. Indeed, this illustrates the other flaw in thinking that we are only being exposed to small amounts at a time. In fact we are probably exposed to dozens and dozens of environmental chemicals in our food every day: pesticide and herbicide residue, various contaminants and byproducts of the manufacturing processes, and additives in the food itself, as well as those in food packaging. Several experts told me that each of us could be exposed to one hundred different kinds of these chemicals a day. What are their combined effects in our bodies? At this point that's anyone's guess. Some scientists are beginning to study combinations of common chemicals that we are constantly exposed to, but this is a new frontier in science. And for their part, the EPA and FDA do not yet consider combined effects.

Generally Recognized as Safe, or GRAS

The other piece of legislation that essentially allows manufacturers to greenlight their own chemicals for use in our foods is what's known as "generally recognized as safe," or GRAS—this represents one of the biggest hurdles for adequate protection of the public health. The provision was originally intended to give common food ingredients added to foods—such as salt, sugar, and vinegar—an exemption from testing and scrutiny before manufacturers could put them in our foods. But what actually happens is that chemical manufacturers use GRAS as a way to fast-track products to market without review for safety by the FDA, from food additives to chemicals for food packaging.

Due to a severe backlog in decision making on new substances coming to the market, in 1997 the agency made the GRAS system officially voluntary, only reviewing the data if the company submits it and only offering a nonbinding opinion about the substance. This means the agency will comment on the substance and may ask some questions, but it will not approve or disapprove of its use. The FDA's position is that the manufacturer is responsible for ensuring the safety of its product. But chemical manufacturers can even choose whether the agency reviews a new substance at all. And if a manufacturer does submit a notification to the agency and is asked a lot of questions about the data, the manufacturer can simply withdraw the notification and use the substance. "Many times, the FDA has plenty of questions about the safety of the substance, in which case, the company can ask the agency to stop reviewing the GRAS notice. They then withdraw it from the system and continue to market it or start marketing it without resolving those problems," Maffini said. "And in some cases the problems are really big and serious safety concerns."[9]

One example is green tea extract. This is a highly purified component from the polyphenols found in green tea called EGCG, yet the manufacturer is calling it the perfectly harmless sounding "green tea extract" for use in teas, sports drinks, and juices (remember my Whole Egg Theory here). But the literature contains troubling findings about this substance, including evidence from human cell studies that it may cause leukemia in fetuses and a rat study showing it affected the thyroid, testes, spleen, liver, and gastrointestinal tract. In some cases the animals were dying of liver tumors. The manufacturer asked the FDA for a GRAS approval but the agency was concerned and asked for more information. Instead of complying with the request, the company insisted the findings were unimportant and withdrew the request, resubmitted it, but then withdrew that one as well. Maffini and her colleague Tom Neltner only discovered this through a Freedom of Information Act request to the FDA. The company using the extract told Maffini and Neltner that they were not selling the product in the United States, but the researchers found more than twenty-five products on grocery store shelves with EGCG as a named ingredient.[10] Maffini said it is probable that manufacturers are calling the extract something else and rephrasing it so that it "sounds very natural" on product labels (in other words, beware of "green tea extract").[11]

What's more, according to Maffini's research, approximately one thousand chemicals on the market today *have never been reviewed by the FDA*.[12] And little to no data exists for chemicals added to our food before 1958, because the agency does not have a complete list of the chemicals that have been grandfathered in. In other words, no one knows how many chemicals are on the marketplace. "Nobody knows because the FDA doesn't have a reporting system," Maffini said.[13]

And concerns remain even after chemicals are reviewed by the FDA. A comprehensive analysis conducted by Maffini and her colleagues found that almost two-thirds of FDA-regulated additives in our food and food packaging have been declared safe without ever being fed to animals in a controlled toxicology study (as faulty as they are). Their review also found that only 21 percent of chemical additives purposely added to food have been researched well enough to reasonably claim a safe dose, that less than 38 percent of FDA-regulated additives are backed by animal feeding studies to evaluate their health effects, and that less than 7 percent of food additives have any data about how they affect reproduction or development.[14] "If you are putting a chemical in a food that

99.9 percent of the population will consume, you would expect it was fed to a rat at a minimum," Maffini said. "We are exposed to these additives from the womb to tomb." Even the FDA's former deputy commissioner for food, Michael Taylor, acknowledged how little the FDA knows about the majority of chemicals added to our foods. "We simply don't have the information to vouch for the safety of many of these chemicals," he told *The Washington Post* in 2014. But when I asked the FDA more recently what it was doing to change its practices and better protect the public, this was the response I received: "Companies are responsible for ensuring that ingredients added to food are safe and comply with the law."[15]

Maffini's review of GRAS notifications also revealed other troubling facts about the scientists the industry hires to perform research on its chemicals. First, they usually are retired toxicologists and professors not attuned to current science, so they will easily dismiss anything that has to do with endocrine-disrupting chemicals. "They definitely do not believe that chemicals in very, very small amounts can actually have a great effect later in your life," Maffini said. In addition, she identified ten people who served on at least twenty-seven or more scientific review panels. One person, now in his late eighties, served on almost 45 percent of the panels that Maffini and her colleagues analyzed. Maffini said that after these scientists approve one chemical for the industry, the manufacturers tap them repeatedly. "So there is this combination of things—you have this small group of people being constantly called to be part of the panels, and they are working on scientific concepts from the 1960s and '70s," Maffini said.[16] According to research conducted by the Center for Public Integrity, some of these same scientists served as consultants for tobacco companies during the 1980s and 1990s.[17]

Good Laboratory Practices

Of course, many independent, academic scientists are attuned to the latest science and do not have industry connections. The scientists I interviewed for this chapter and for chapter 6 are among the best in the field and there are many more, but the EPA and the FDA do not consider their findings when making decisions about the safety of environmental chemicals for yet another reason: academic labs do not follow a protocol known as "good laboratory practices," or GLP.

The concept of GLP goes back to the mid-1970s when the government investigated Industrial Bio-Test Laboratories, which operated the largest toxicology-testing facility in the United States during the 1950s, '60s, and '70s. Its clients included many of the major manufacturers, among them DuPont, Dow, and Monsanto. In 1977 a government investigation revealed extensive fraud in its practices that affected scientific findings about products widely used in the home or in agriculture, including Roundup, that herbicide in your Ben & Jerry's, and PCBs, those now-banned chemicals that persist in the environment and remain pervasive in our food. The fraud in this lab was so extensive that some manufacturers were citing studies conducted on animals that did not even exist, the researchers were substituting new animals for those that had died during testing, and they were not even feeding animals the test substances.[18] As a result of this scandal, the government mandated that when industry uses its own laboratories to conduct studies of its products, it has to abide by a set of standards, known as good laboratory practices. This sounds like smart policy: create a set of standards for all scientists to follow in their research. And initially GLP was designed to stop fraud with meticulous record keeping and strictly enforced guidelines. However, these guidelines are prohibitively expensive for smaller academic labs to follow, even though this is where most cutting-edge science is taking place. Academic labs like those of Blumberg, Gore, and Zoeller simply cannot afford or don't have the staff to adhere to GLP. "We don't get that much money on our grants to do it, and our universities are not set up with animal facilities that can record every time someone enters the animal facility or leaves and what they are doing at that time—so my work cannot be GLP compliant. Which means that the FDA and EPA can't look at it," said Laura Vandenberg, another academic researcher who studies endocrine disruption.

The trouble is that GLP does not address scientific principles or the design of a study. "All that GLP means is that the animals existed, that the data was collected, and that you know who was there when it was done—it does not mean that the study was designed properly, it does not mean that it was conducted with good scientific principles in mind; it just means you did record keeping," Vandenberg said.[19] In some cases the protocols are based on the *Redbook*, an FDA publication that is like a cookbook for testing chemicals. The problem with the *Redbook*, which first came out in 1982 and was revised in 1993 and 2000, is that its framework is old, Maffini said. "It doesn't account for contemporary science and knowledge. Science is not a fixed target, science moves."[20]

The evolving science of chemical exposures must take into account low-dose exposures and sensitive end points because these findings reveal the effects of the chemicals in our food supply on the various systems of our bodies—and on the most vulnerable among us, particularly fetuses, babies, and children. Maffini points out that this failure is often not the fault of the toxicologists themselves but a result of the outdated guidelines from the regulatory agencies that the toxicologists adhere to. "They don't make decisions on the doses, the doses are given to them," Maffini said.[21]

Ultimately this means that the work of the academic scientists—who are working to protect the public interest—is completely ignored by the regulatory agencies.

Many Chemicals Are "Grandfathered In"

Bisphenol A (BPA), the ubiquitous chemical added to much of our plastic and canned food packaging, was approved for use in 1963, and its FDA safety assessment from that time still stands. Here's an example of how the legal framework for food additives handcuffs the agency: Once a food additive is approved for use—whenever that was—food manufacturers can use the substance without notifying the agency. And in the case of BPA, hundreds of formulations of BPA epoxy resins are in food packaging and have different characteristics. Manufacturers are free to use any of these without notifying the FDA about the nature of these formulations. This obviously makes little sense from a public health perspective because both the science and our diets are always changing. As of 2017, one thousand studies showed harm in animals and one hundred studies showed harm in humans from the chemical—much different than what was available when it was approved for use in 1963. Plus, our use of plastics and other packaging has increased dramatically since that time—recent estimates are that consumers buy one million plastic bottles every minute worldwide. And while BPA is leaching into our foods and drinks directly, its use in the manufacturing process and the disposal of that plastic (much of which ends up in the oceans and then in seafood) contaminate our environment and can find its way back into our foods where we might not expect it. To the FDA's credit, in 2010 it acknowledged that, given the health concerns of BPA, the agency should be able to apply "the more modern framework . . . for oversight of BPA than the current one."[22] However, that still hasn't happened.

In 2008 thirty independent scientists wrote a paper after reviewing the differences in industry-funded papers versus publicly funded papers (studies conducted by academic scientists and funded by government grants) on BPA research. They found that the FDA and the European Food Safety Authority determined the chemical was safe based primarily on two industry-funded studies and ignored hundreds of independent studies that found an array of health risks associated with the chemical. The authors state that the two studies had "serious conceptual and methodologic flaws" and that the regulatory agencies "have mistakenly assumed that good laboratory practices yield valid and reliable scientific findings (i.e., 'good science')."[23]

The regulatory agencies currently have no reassessment system in place for reviewing many long-existing chemicals. "Once a chemical is considered safe, it is considered safe forever," Maffini said. "If there's a chemical that I'm using in 2005, but I'm using data from 1942—I don't like that. We should be doing more screening, more predictive models, more in-vitro testing, and then, based on all the data we collect, we decide if we need to use animals to do more testing. We didn't have all those opportunities forty or fifty years ago, [but] now we do—why not use them?" Maffini also pointed out that because our diets have changed significantly from even forty or fifty years ago, the safe amount of exposure to a particular ingredient or chemical in our foods is probably much different. "We are now likely eating way too much of something—if you are eating more than the safe level, you are now at an unsafe level," she said.[24]

In addition, manufacturers will often take data on a related chemical and claim the new chemical will behave similarly to the one already tested. But that data could be quite old. Phthalates, those plasticizers researchers found in your mac and cheese, are a good example. Some, like di(2-ethylhexyl) phthalate, or DEHP, were approved before 1958. Today thirty types of phthalates are approved for use in contact with food—in many kinds of food packaging, and all along the food-manufacturing process—and more than half have no toxicity or exposure data.[25]

Pressures Within the Agencies

When scientists at the FDA or the EPA do conduct their own research, several independent scientists told me that the agencies' scientists are pressured to come up with findings that support the agency's current stance on particular additives. Frederick vom Saal, the University of Missouri professor of biological sciences who studies BPA, has found that regulators and scientists within the FDA are pressured to support its position that the chemical does not pose health risks. Vom Saal co-authored a paper that argues that one FDA scientist contradicted his own data in the conclusion of the paper in order to conform to the agency's stance.[26] According to vom Saal and his co-authors, in the 2010 paper the FDA scientist Daniel Doerge reported significant changes between infant and adult rhesus monkeys in the metabolism of BPA, which would confirm earlier findings in rodents that infants have a harder time metabolizing and therefore excreting the chemical from their bodies than adults. Yet in the conclusion to his paper, Doerge and his co-authors claim that there was no evidence for diminished metabolism in infant monkeys.[27] Vom Saal attributes this contradiction to pressure from regulators to "not find anything that disagrees with the FDA stated position that BPA is safe."

Contradictions between the findings in a paper and its conclusions are typical of industry-funded studies, and now of FDA studies too, vom Saal said.[28] A number of scientists called for a formal investigation of Doerge for scientific misconduct, but the FDA refused, saying this was just an example of scientific disagreement.[29] For his part, Doerge said that vom Saal and the others' claims are unsupported since his research for the paper underwent "extensive peer-review" as well as review by panels of experts, including those at the FDA and the European Food Safety Authority. And Doerge stands by the conclusion of his paper, which says that the effects of BPA on rats overpredicts the effects the chemical could have on primates.[30] While the FDA says this is a matter of scientific disagreement, this debate between scientists gets to the heart of how BPA should be regulated, particularly when infants are exposed to the chemical.

Zoeller told me that he saw pressures within the agency while working at the FDA as well. "The FDA is conflicted. The regulatory arm at FDA has an agenda that the data don't always fit into, and instead of following the data, they follow their agenda," he said. In the case of BPA, the regulators at the FDA are employed by the same agency that is conducting the science, and "that creates the potential for very significant conflict of interest," he said.[31]

The question then becomes what the agency's agenda is based on. Zoeller said a prime culprit is the culture that has developed at the FDA. He called it a turf issue and said many people there resent having someone come in to evaluate their work. While Zoeller was working at the FDA, he recalled, one employee said that no one at the agency had ever made a mistake. "In the forty years that he'd been working there, they never made a mistake? I'm going, 'Oh, my god . . . it's an ego thing,'" Zoeller said. Maffini noted that a scientist who has worked at the FDA for forty years is going to find it difficult to acknowledge something he once said was safe may not be. Regulators are also always being criticized, said Zoeller, either by industry or by environmentalists, or both. "That pressure from the outside may also support a bit of a bunker mentality. I think the consequence is that they interpret data from a fundamentally biased perspective," he said. "I have to admit my experience with FDA food safety spawned this kind of thinking. They were actively resistant to their own data concerning BPA, and, as such, they functioned more like a product defense organization than a regulatory agency." Zoeller said this culture allows the industry to apply pressure effectively—and there is no doubt that the industry does apply pressure. Ultimately, Zoeller said of his experience at the FDA, "I never saw any evidence that they are motivated by public health protection."[32]

Zoeller also took note of what he called the turnstile effect. By this he meant the revolving door between people who leave the regulatory agencies to work for industry and vice versa. One of the most widely reported examples is Michael Taylor, who stepped down in June 2016 as deputy commissioner for food at the FDA; before joining the FDA he had been the vice president for public policy at Monsanto.[33] There are also many lower profile examples that nonetheless have profound effects on regulation outcomes and public health protection. In 2011 the firm Steptoe & Johnson, which lobbies for and defends the chemical and food industries (among many other industries), hired Mitchell Cheeseman, who had served as director of FDA's Office of Food Additive Safety, and Ralph Simmons, who was Cheeseman's senior adviser at the FDA. They joined the firm as managing director and partner, respectively.[34] "There's a revolving door—all these senior people from the Center for Food Safety and the FDA over the last seven years or so are now working for K Street law firms," vom Saal said. "And they're not making little bits of money."

Zoeller said it is not uncommon for people to stay with the

EPA long enough to get a decent retirement package before jumping to industry for double the salary plus retirement benefits. "There are actually many examples of former EPA employees going to industry," Zoeller said.[35] In 2017 President Donald Trump nominated a well-known friend of industry, Scott Pruitt, to head the EPA. Within his first month at the agency, Pruitt denied the petition to ban the common pesticide chlorpyrifos, which laces our food supply and has been shown to cause brain damage and reduce IQs in children, and is linked to lung cancer and Parkinson's disease in adults. Documents received through a Freedom of Information Act request revealed that Pruitt met with the CEO of Dow Chemical, maker of the pesticide, just before he rejected the ban.[36]

Ultimately these agencies are not protecting our health, and we cannot rely on or trust their regulatory decisions. The trouble is that so many of these chemicals pollute our larger environment, not just our foods. So even if you avoid all packaged and processed foods (a near impossibility), you are still exposed to these chemicals in our environment every day. In addition, most Americans are simply not aware that these kinds of regulatory failures allow so many untested chemical combinations into our foods and our bodies.

This brings up another major issue: consent. When you buy a can of green beans, for example, are you consenting to eat the BPA or the other additives in that can? If there is no indication on the package that you will be consuming these chemicals with your food, can the food industry claim that you are truly making a choice? Laura Vandenberg put it this way: "There is no consent in typical environmental exposures; when you eat food out of a can, you might be consenting to eating the green beans—but you're not even thinking or imagining that you are being exposed to a dozen other chemicals, the can lining, the can, and the preservatives and all that other stuff." So the food industry's claims of choice are false—and proponents of free market capitalism who oppose regulation would say that if you are concerned about these chemicals, you are free to buy foods that do not contain them. But how can you know they are present if they are not listed on labels? And because the regulatory agencies are not testing final products for environmental chemicals, even they don't know what's in them, never mind requiring that they be labeled.

Kim Richman, a lawyer who works on consumer rights and litigates consumer protections, said that consumers have few protections under the law. In fact, Richman and other consumer lawyers need to be able to catch a company making a false advertising claim about its product in order to file a lawsuit with any chance of success. In May 2016, Richman and his colleagues initiated a class action suit against Quaker Oats for selling oatmeal that contained trace amounts of the herbicide glyphosate, or Roundup, while labeling the product 100% NATURAL.[37] The industry's defense in such a case typically rests on the regulatory agencies' allowance of trace amounts of this herbicide in our foods. Which brings us back to the beginning—so long as the agencies continue to rely on the outdated toxicology model of the dose makes the poison, incomplete and inadequate scientific testing will be the basis for decisions that allow trace amounts of chemicals in our foods—and the basis for the industry's continued defense of their use.

What's more, industry is protected by proprietary information, which allows it to keep information from the public. For example, I contacted Gerber about its use of plastic cups in dispensing baby formula for a product called BabyNes. I was concerned about running hot water through these small plastic cups (they are like K cups for coffee machines) and asked what the cups are composed of, in addition to the polypropylene that Gerber lists on its website. Polypropylene is a type of plastic that requires the use of antioxidants and additives to be stable, and I wondered whether the company could be exposing newborn babies to various endocrine-disrupting chemicals as a result. The response from a company representative was telling: "The capsule used for packaging BabyNes infant formulas is made out of polypropylene. . . . The exact composition of our capsules is proprietary information. Polypropylene is safe for packaging food and formulas for infants." The Gerber official added that its infant formula and packaging comply with all relevant FDA requirements. In other words, we have to take the manufacturer's word that its product is safe for babies without knowing all the ingredients. And because companies provide their own safety data to the FDA, saying a product is "FDA approved" is hardly a reassurance.

Clearly the laws are in place to protect industry and not the public. "It is interesting because the chemical is almost being viewed as having a certain right," Zoeller said. Zoeller, who has worked for both the FDA and EPA, told me the following, which perfectly illustrates the fundamental problems with the way chemicals are evaluated:

> The "risk-based" approach that the EPA uses is based on the concept that the dose makes the poison. The assumption underlying this is that there is a dose threshold of toxicity, below which no harm occurs. Because the pesticide manufacturer works with the agency to develop and interpret safety data, it is fair to say that risk assessment is a collaboration between the regulators and the regulated. Because the goal is to use toxic chemicals that people will be incidentally exposed to, it is fair to say that the goal of risk assessment is to expose the human population to toxic chemicals safely.

The bottom line is that we do not truly have a choice when it comes to what goes into our bodies but rather the industries and the conflicted regulatory agencies decide what we will be exposed to—based not on rigorous science but on what is most profitable for industry and the economy in general. We must ask ourselves how long we will allow profit and wealth to supersede human health concerns. What are the benefits of these chemicals versus their detriments? Food manufacturers would argue that food additives and chemical-laden packaging extend shelf life, keep food production costs down, and enhance flavors; chemical manufacturers would argue that their various pesticides and herbicides protect crops and help farmers. But at what cost? "The public has the general conception that the EPA and FDA are working very hard to protect them, and the reality is they are working very hard to protect the interests of the chemical industries and the drug industries," Blumberg said.

It's about time we realize that we cannot balance productivity, efficiency, and profit with the health of people and the protection of the public. "When you allow a system to develop where every baby born in this country has literally well over 100 manufactured chemicals in their body—most of which have never been tested," Zoeller said. "You have to ask, what does that say about our society?"[38]

PART FOUR

Understanding Our Food Culture and Creating Real Change

We cannot talk about improving our food system or public health without improving the very structures of our society. Overwork, longer hours, lower pay, lack of paid parental leave, racism, sexism, and class and cultural divisions all factor into our deteriorating health. As chapters 6 and 7 make clear, we have little food autonomy—we have no say about what goes into our foods, and in many cases we are at the mercy of Big Food and Big Ag's decisions about what we will eat. The idea of choice when it comes to our foods is an illusion.

Whether we eat well ultimately has little to do with individual choice; rather, how well we eat is affected by four key issues. The first two are the related issues of money and time—with declining wages and longer working hours, many people simply don't have the money or time to purchase whole foods and cook. The third is the skill; cooking from scratch is something that has to be learned, and for many in the past one or two generations, this skill has been completely lost with the prevalence of processed and fast foods in the home. The fourth key issue is the attitude of the culture in general, which encourages the consumption of fast food, inexpensive convenience foods, and otherwise poor-quality meals, snacks, and drinks. Many of these issues break down dramatically along race and class lines, with poor people and people of color often affected the most by our problematic food culture. The issues are complex and often hard to address; the result has been that many in the food movement avoid these tough conversations, either ignoring the subjects entirely or reverting to industry's refrain to choose better, healthier foods. But this is inadequate. If we really want to tackle our current food and health crises, we will have to address the thorny issues of race, class, and gender as they play out in our current food landscape.

One of the most powerful aspects of food is that it can be a great unifier. If enough of us agree that the status quo is no longer acceptable, we can be a powerful force. But first we must know exactly how the industrial food system has usurped our food autonomy, undermined any real notion of choice, and worked for decades to get all of us to accept its products as the basis of our diets. Only then can we work toward solutions that will amount to real change.

Food Choice

The Culture of Our Upside-Down Food Environment

Consider the strange fact that it costs far more to purchase foods that have fewer ingredients—foods that are free of environmental chemicals like additives, pesticides, and herbicides, and organic animal products raised without growth hormones or antibiotics. All this costs farmers, producers, and ultimately consumers more. Indeed, buying foods without these added ingredients means that you pay a premium. This is not logical. The popular political comedian and talk show host Bill Maher raised this point when he interviewed President Barack Obama and brought up "food purity": "Somehow it got to be elitist or liberal—I don't get that—to want food that isn't full of pesticides, or food that wasn't created by torturing animals in factory farms, isn't full of antibiotics . . . I feel like we are upside down on food," Maher said.[1]

We *are* upside down on food, a sure sign that Big Food and Big Ag have done a masterful job of changing cultural attitudes about it. It's also an indication of how deeply industry has influenced regulation, food policy, food pricing, and so-called choice. After all, food culture and food policy are a two-way street of influence. The food industry's chief narrative is that grocery store shelves offer us endless options—from junk food to healthy, organic, and natural products. But this narrative is completely false. First, the consolidation of our food supply by a handful of corporations means that the majority of these choices are controlled by the few. These companies decide what our options are and what the content of those foods will be. Ten companies control nearly every large food and beverage brand in the world: Nestlé, PepsiCo, Coca-Cola, Unilever, Danone, General Mills, Kellogg's, Mars, Associated British Foods, and Mondelez.[2] And the majority of those foods contain what are essentially variations of the same ingredients: wheat, corn, soy, sugar, sodium, vegetable oils, and an array of food additives—and these are heavily subsidized. This creates a false pricing structure, in contrast with produce, tree fruit, and nuts, which are not subsidized. And when it comes to seed, which is the foundation of our food supply, just six companies control 98 percent of the world's seeds. Our meat and dairy have been consolidated too: four companies slaughter 81 percent of all beef and control 70 percent of all milk sales.[3] Even organics have been consolidated with just a few companies controlling the vast majority of organic produce and organic milk.

But the food, agricultural, and chemical industries leave us with little choice in other, more subtle ways. These poorly regulated industries allow for the use of ingredients of questionable safety and quality, which undermines the claim that we are fully free to choose what we eat as informed consumers; indeed, many foods and drinks are not even properly and thoroughly labeled with the information we need to understand what we are consuming.

When it comes to protecting our food supply and keeping it safe, the failure to institute smart and effective regulation also means that certain segments of the population are harmed more than others. Free-market proponents claim that more regulation would place a bigger burden on low-income consumers who rely on low-cost foods, but these claims don't take into account the fact that they—and all of us—are already paying an enormous health burden due to the failure to properly regulate the industries in both their free rein on advertising and, even more importantly, their use of environmental chemicals. Those who rely on cheap processed foods for a large percentage of their diet are exposed to more environmental chemicals than people who eat higher-quality foods, which tend to have fewer industrial additives. One study found that people living in underserved areas are more frequently exposed to bisphenol A, or BPA, the common food-packaging chemical, than people in the general population. This chemical, along with a slew of others, is found in the lining of cans and in plastic containers, and BPA has been shown to leach into food and contribute to obesity, diabetes, and a host of other health problems. This is also at least part of the reason that rates of diet-related disease break down along economic lines. Poor people have nearly double the incidence of diabetes than those in the highest income bracket, 10.1 percent and 5.5 percent, respectively. When it comes to food and health, economics matters greatly; the more money you have to spend on food, the higher quality food you will buy.

Those who live in food-secure households spend 30 percent more on food than those living in food-insecure households.[4]

Poverty plays a huge role in understanding the problems with our current food supply and who is harmed the most by this system. In 2012 forty-seven million people were living below the poverty line in the United States, and fewer than 10 percent were white. African Americans made up 27 percent of the poor, Native Americans 26 percent, Latinos 25.6 percent, and Asian Americans 11.7 percent.[5] This means that high numbers of people of color are food insecure, not knowing where money for future meals will come from while also relying on cheap convenience food for meals that provide more caloric bang for the buck.

The fact is clearly apparent in the class- and race-based health disparities in this country. While obesity and diabetes are prevalent among all racial and ethnic groups, some groups are affected more than others. White women have a 31 percent incidence of obesity, Mexican American women have a 41 percent incidence, and black women a 51 percent incidence—one in every two African American women. Diabetes also affects black women more than any other group. Their incidence of diabetes is 11.6 percent, which is more than double the rate for white women at 5.8 percent; Hispanic women have an incidence of 9.4 percent.

On top of the largely unregulated food supply itself, the advertising industry essentially has free rein in marketing, especially to children and communities of color, which are specifically targeted. Some of the most egregious advertising tactics are aimed at young children. One study found that toddlers can identify the golden arches of McDonald's before they have even learned the letter *M*. After looking at more than one hundred brands, the researchers also found that children prefer foods with familiar logos and that these foods are high in sugars, poor-quality fats, and sodium. The study also found that seeing a logo stimulates the pleasure and reward regions of children's brains—the same regions that function in various types of addiction, including drug abuse.[6] When manufacturers get children, babies, and even fetuses in the womb accustomed to processed foods from the start, the industry is developing lifelong customers.

In a comprehensive analysis of fast-food nutrition and marketing conducted in 2012 by the Rudd Center for Food Policy and Obesity, researchers found that the food industry specifically targets teens and minority youth more frequently and with more harmful foods than any other group. The researchers found that black youth saw 50 percent more fast-food ads on TV than their white counterparts, and McDonald's targets Hispanic children, especially preschoolers, with ads on Spanish TV. Two of the same researchers reaffirmed these findings in 2016 with another study that found that black youth viewed 50 percent more ads than white youth of the same age. And it is not because black youth watch more TV than white youth, which was long considered part of the explanation for this discrepancy. Rather, the researchers found, junk food marketers are specifically targeting black youth on the networks they are more likely to watch—Fuse, Nick-at-Nite, BET, and VH1—which are also the networks that air the most food advertising.[7]

Then there was the 2013 marketing marriage of the mega pop star Beyoncé and Pepsi. This merger is a prime example of the predatory nature of Big Food (or in this case Big Drink) corporations. Beyoncé, the reigning global superstar, signed a $50 million deal in 2012 to hawk a product that is well documented as leading to obesity and diabetes, with particularly bad effects among African Americans and Latinos. Yet the industry continually ignores the health risks for their consumers and instead presents its products as choices to be consumed "in moderation." Adding to the contradictory messaging was Beyoncé's involvement with Michelle Obama's Let's Move! campaign, created to address childhood obesity. So, when a young girl looks at someone like Beyoncé, she might think that she can drink Pepsi as long as she dances, exercises, or otherwise moves enough. This is straight out of the industry playbook and is part of its personal responsibility messaging: exercise more and eat less. This relies on the false and outdated "energy-balance" model—or calories in versus calories out. Plus, remember that the hormone-altering chemicals that we are exposed to in those soda cans and plastic bottles likely play a role in weight gain and metabolic disorders. The knowledge we have about race and class health discrepancies, the targeted advertising to these same groups, and the prevalence of poor-quality foods in many of these communities paints a pretty clear picture as to who is really to blame.

And here we come back to the issue of our upside-down food culture. The food industry and its allies in the agricultural and chemical industries prey on the vulnerable by pumping out barely regulated products at low cost and then marketing them with insidious advertising campaigns. These ads normalize their products and simultaneously do one of two things, depending on which one produces better profit margins: they try to create a monolithic American food ideal to be accepted by all, or they perpetuate and antagonize racial and income divisions to market their products to different groups with different sensibilities. These industries have become alarmingly adept at navigating the slippery terrain of our upside-down food culture. Or maybe not—after all, they are the ones who orchestrated it from the beginning.

In the first instance, the food industry has been effective at creating a shared food sensibility—it has been masterful at convincing the American people to buy its highly processed packaged foods and fast foods because they represent "American" values. As a result we have universally accepted industrial foods into our homes, and most Americans have bought into the food industry's messaging about health and nutrition. This acceptance cuts across all demographics—as a nutrition educator I've worked with people living modestly and struggling to afford better-quality foods as well as multimillionaires living on Park Avenue in Manhattan, and I hear the same misunderstandings about food, nutrition, and health over and over again. For example, people widely believe that eating low-fat dairy is healthful, cereal with skim milk is a reasonable and healthy breakfast, they should avoid butter and other saturated fats, industrially produced sliced "whole-wheat" breads are healthy, organic chips and crackers are healthy options . . . this list could go on indefinitely, but you get the point (I also see the same processed foods in all these kitchens).

What's more, even those who would be considered food elites, the well-educated liberal types living in progressive enclaves like Park Slope, Brooklyn, or Berkeley, California, are still eating processed junk food with regularity. I've lost count of how often I've seen products like Raisin Bran, Kashi cereals, low-fat strawberry Stonyfield yogurt, skim milk, or Annie's Macaroni & Cheese—all of which their manufacturers tout as healthy but are not—in the kitchens of friends and neighbors here in Park Slope. As a member of the Park Slope Food Coop, I am privy to the buying patterns and mind-sets of people looking to feed themselves and their families well but who are all too often swayed by claims on packages and the convenience factor. I see parents buying their children organic versions of Cocoa Puffs, Pop Tarts, and Goldfish crackers—and paying significantly more for them—without understanding that these are essentially the same poor-quality foods fancied up and given a health halo.

And even at the height of the Occupy Wall Street movement, when I was surrounded by self-proclaimed radicals protesting corporate influence and power, I saw Occupiers consuming the ultimate corporate foods and drinks: veggie sandwiches from Subway, vegan oatmeal from McDonald's, or Coca-Cola. This indicates not just how pervasive these corporate foods are, but how our consumption patterns—with food in particular—fly under the radar about what is acceptable to eat or reject, not only from a health perspective but from a political perspective. The Occupy example is partly a reflection of how difficult it is to eat healthy, noncommercial foods when we are away from home, but it also demonstrates how industrial food products have become so normalized in the culture and achieved such near-universal acceptance that no one questions them. For these reasons it is crucial that we begin to view food through a political lens.

Industrial food products are reaching into every American home. Although fast food was originally created as a way that "working-class families could finally afford to feed their kids restaurant food," a recent study revealed that it's now the American middle class that buys the most fast food. This study found that a household earning $60,000 a year eats the most fast food, and one earning $80,000 is actually more likely to eat fast food than one earning $30,000. Findings from the Centers for Disease Control and Prevention make a related point: nearly half of obese adults in this country are not poor but middle-class, earning at least $77,000 for a family of four. The reasons for this are multifaceted—they are physiological for sure (for example, the development of taste preferences in the womb and after birth, or the disruptions to our microbiotas), but they are also psychological: the food industry has succeeded at normalizing highly processed convenience and fast foods so that most Americans think that Frosted Flakes, Doritos, Oreos, a Big Mac, fries, and Coke are perfectly fine foods for regular consumption. And perhaps even more insidiously, the market for so-called healthy organic processed food continues to grow with ever more consumers falling prey to this new kind of deceptive marketing. Identifying with an American sensibility vis-à-vis food choice has worked particularly well in persuading immigrants who want to assimilate into American culture to leave their healthier food traditions behind. A 2011 study of Latina immigrants found that lower consumption of traditional food products was associated with unhealthy dietary habits and obesity. The researchers urged immigrants to maintain traditional healthy foods and keep them available in their communities.[8]

In other instances, when it suits industry's bottom line, it creates rifts in a nation already divided by race and income. The food industry has been capitalizing on and perpetuating these differences for decades. And the end result is a food culture that now considers whole, less-processed foods elitist and denigrates those who teach people about the benefits of these food as part of the "nanny state" and decidedly un-American. A 2015 McDonald's commercial presents a clear example of this. The commercial addresses "vegetarians, foodies, and gastronauts" and tells them to "kindly avert your eyes." The viewer then sees an extreme close-up of McDonald's signature Big Mac as the narrator says, "There is no quinoa or soy, Greek yogurt, or kale to be found here," adding, "and while it is massive, its ego is not." This ad essentially says, "Forget all those food snobs, food movement elites, and otherwise obnoxious kale-and-Greek-yogurt-eating people of the upper classes. McDonald's makes food for the masses—the 'real' Americans, the hard-working folks without ego or pretense."[9]

The food industry's defensive posture is that to attack the American way of eating—which the industrial food system obviously represents—is to attack the American way of life. This posturing mirrors precisely the narrative about "coastal elites" that ridicule and scold "regular Americans" for their choices and values. You'll remember "Joe the Plumber" and Sarah Palin's "soccer moms" from the 2008 presidential election campaign as tropes used to invoke the "we're one of you" messaging of the Republicans. And somewhat surprisingly Donald Trump—a billionaire—used this same tactic successfully in his 2016 campaign.

Back in 2010 Palin made news when she went to a school in Pennsylvania to hand out cookies to children as a rebuke to the Pennsylvania State Board of Education, which was encouraging schools to limit sweets and provide healthy foods. At the time, Palin said she brought cookies to the school to symbolize her opposition to "a nanny state run amok." Her message was loud and clear: those advocating for healthier foods were somehow depriving children of their God-given right to eat sugar-laden, processed foods.[10] And then in 2016 Trump used his love of fast food as a powerful symbol of his identification with so-called regular Americans, although in the iconic photo of him eating a KFC meal, he is sitting on the plush white leather seats of his private jet.

Another prime example of the alleged nanny state was the proposed large-sized soda ban, championed by former New York City mayor Michael Bloomberg as a means of combating rising rates of obesity and diabetes. The reaction to his campaign brought to the fore issues of race, class, and choice that had largely been masked by the food industry. In this case, the American Beverage Association (ABA), a lobbying group for corporations like PepsiCo and Coca-Cola, campaigned against the soda ban through the language of choice and a thinly veiled appeal to racial and class politics. The group used the term *choice* as a way of pitting the average American against elites like the billionaire mayor Bloomberg. Racial politics was another undercurrent in this campaign, because the beverage industry made a strong appeal to Latinos and African Americans by painting a picture of a paternalistic government that claims to know best—and Bloomberg, a wealthy white businessman, was the perfect poster child. But what the industry really did was flip elitism on its head—that is, the ABA represents the actual elites, yet it convinced average Americans that by defending billion-dollar, multinational corporations, they were standing up to elites and protecting civil liberties.[11]

It's important to understand how this narrative has been flipped on its head. In the early days of processed foods, industry advertisements targeted middle-class women—who had the money to spend on processed products—by touting their convenience and, even more important, the message that these new foods could buy higher social status. The food industry promoted using cake mixes and sliced supermarket bread, buying boxed cereals, and serving TV dinners as modern and advanced ways to feed the family—and at the time, these cost more than simple, whole foods. The ways of the past, such as growing our own vegetables, baking our own bread, and cooking just about everything from scratch, were portrayed as old fashioned. But in the 1970s Big Food suffered a backlash among segments of the population when the more critical generation of baby boomers demanded better, healthier food.[12] You might call this the beginnings of the food movement in America, with a push for higher-quality, purer, and sustainably produced foods. Indeed, the Park Slope Food Coop started in 1973 with just a handful of members and a tiny space as a response to the industrialization of our foods.

But more recently the industry has succeeded at framing the increasingly popular healthy food movement as elitist and even un-American with its divisive marketing. Food has always been a marker of class and rank in any society, writes the food historian Felipe Fernández-Armesto: ". . . class differentiation starts with the crudities of basic economics. People eat the best food they can afford: the preferred food of the rich therefore becomes a signifier of social aspirations."[13] But in recent years the food industry has managed to subvert this long-standing food trend with brilliant marketing.[14]

And now that two or three generations are acculturated to industrial foods, we have collectively forgotten what real food is. As a population, we have dietary amnesia. Children especially have accepted our industrial food world and in some cases are the best ambassadors for its products through the age-old tactic of peer pressure. I worked with an overweight seven-year-old-girl who struggled with this very issue. Her friends were mostly the children of working-class parents, and she refused to bring the healthier items that her parents packed for her school lunch, saying that the other children teased her for having carrot sticks with hummus, turkey roll-ups with avocado, and a canteen of water, while they lunched on brightly colored packages of Doritos, plastic-encased Lunchables, and cans of Coke.

Anyone who has children—or who was a child for that matter—will remember the importance of having certain brand-name goods in order to fit in. With the exception of some of the crunchier communities that wholly reject consumer culture, no one is immune to this type of pressure in American culture. Having Doritos and Coke for lunch is not much different than wearing the latest Nikes to school—they are symbols of having enough money to afford the brand names and not be seen as uncool by, for example, eating unbranded foods or wearing knockoffs. The only foods that are not branded are whole foods, which do not have labels and are not marketed as "American food" for average Americans—that is, part of the mainstream. And even though corporate food products are now cheap—much less expensive than organic whole foods in many cases—industrial food has retained a certain prestige associated with their brand names.

All these complex and confounding factors illuminate exactly why the ideas that the "food movement" often promotes—that access and education are the remedies for race- or class-based health disparities—have failed to address our national food and health crises. It long ago became a truism that the health of low-income people of all races, and in locations both rural and urban, suffers because they don't have access to healthier foods and live in "food deserts." While lack of access does present one obstacle to healthier eating, it is but a small part of a far more complex picture. The oversimplification of the problem of poor eating habits and poor health and the emphasis on access belies the real power issues at work. And just as the industry has presented choice as a way to absolve itself of any responsibility for the poor quality of its food products, food movement leaders have presented choosing to buy organic and healthy foods as the solution to the problem of poor health. The latter recommend "voting with your fork" or making better choices. But this simplistic formulation is inadequate. Food movement leaders point to racism in the food system but ignore the many ways it parallels the racism, classism, and paternalism that sometimes surface in the food movement itself.

When the food movement leader Michael Pollan says, "Eat food. Not too much. Mostly plants"—his most famous catchphrase—he and others like him fail to acknowledge that doing so is an exercise of privilege and power in more ways than just economics and access.[15] The vast majority of school-age children do not eat healthy school lunches designed by Alice Waters, as kids do in Berkeley, California, where real food is the norm. And not only do the advertisers and the food industry capitalize on a lack of quality food options, low incomes, and lack of time, but they also capitalize on this shared sensibility among children and other vulnerable groups who don't want to be further marginalized by their food choices. They want to eat what most other people are eating. This is another crucial yet overlooked way that being able to choose to eat organically and avoid processed and fast foods has become a privileged decision.

When *60 Minutes* profiled her in 2009, Alice Waters said: "We make decisions every day about what we're going to eat, and some people want to buy Nike shoes—two pairs—and other people want to eat Brock's grapes [a special variety from a farmer near San Francisco] and nourish themselves. I pay a little extra but this is what I want to do."[16] Her comment shows just how out of touch many in the food movement are when it comes to issues of income, race, and so-called food choice. She's arguing that spending four dollars a pound for grapes is morally superior to spending money on a pair or two of Nikes. Telling people how to spend their money is not the solution—this is precisely where the sense that the food movement is elitist or paternalistic comes from.

This tactic has historical precedent in its racist and classist undertones. "The idea that frugal consumption is in any way a liberation is as old as capitalism, and comes from the capitalists who always blame the worker's situation on the worker. For years Harlem was told by head-shaking liberals that if Black men would only stop driving Cadillacs . . . the problem of color would be solved . . . [yet] that Cadillac was one of the few ways to display the potential for power." Selma James, the cofounder of the Wages for Housework Campaign, wrote that in 1972.[17] In the relatively powerless situation that many of us find ourselves in, buying a pair of Nikes or even a nice car does indeed allow us to display some measure of power. Yet in many cases those in the food movement do not understand this and instead scold people for their choices. How many times have we heard that if people would just spend less on their TVs, cable service, iPhones, or cars, they would have more money to spend on healthy food? Such paternalism makes clear that these food movement leaders never have to make the kinds of choices that they are telling others to make.

There's another way that others in the food movement sometimes fail to recognize the struggles of the poor and working class when it comes to choosing how to eat. A common trope in the food movement is to demonstrate how whole food ingredients are less expensive than fast food or other convenience foods. This may in fact be true in certain cases; however, as the journalist Tracie McMillan points out in her book *The American Way of Eating*, when she attempted to live on $8.10 an hour while working at Walmart, stocking up on a pantry of whole food ingredients was simply not possible. "My strategy of stocking up on bulk groceries in the first couple of weeks now strikes me as less of an example of thrift than of the truth contained in Gloria Steinem's observation that 'planning ahead is a measure of class,' by which she meant not poise or erudition but income," McMillan writes. "I may save money in the long run by buying dry beans and whole poultry, but it's left me so cash-poor that I'm unable to cope with basic expenses."[18] Furthermore, the time it takes to shop for and prepare that meal is not accounted for in the cost. This often means that time-strapped and cash-poor working-class and middle-class parents are heavily reliant on industrial food products, thereby furthering the normalization of them in average households and school lunch boxes across the country.

At the same time, many of these food movement leaders pay lip service to race- and class-based health disparities, often citing statistics. While it is useful to have an empirical knowledge of statistics that highlight gross inequality in health outcomes (as I outlined earlier), it's equally important to acknowledge that when we pathologize health outcomes, particularly among people of color, we may in fact sabotage efforts to solve the problem or change the narrative. For example, a common belief is that eating clean, healthy, or organic is a white middle- and upper-class phenomenon, which positions people of color solely as consumers of poor-quality junk foods. This narrative frames whites as active, informed consumers while casting people of color as victims of our industrial food system. This results in simplistic ideas about who is healthy or not and who does or doesn't care about their food. It also casts racial groupings as monolithic, promoting some vague notion that all whites or all blacks can be easily lumped into categories with shared sensibilities and desires.

All too often food writers and activists will cite racial disparity statistics and then leave it at that—allowing the thornier issues of how and why to remain unanswered. But recall the findings of the Centers for Disease Control and Prevention that said that nearly half of all obese adults are middle class, not poor, and that those earning $80,000 a year were more likely to buy fast food than those earning $30,000 a year. At the Park Slope Food Coop, which has seventeen thousand members, there is an incredibly diverse population (by all measures: race, class, age, income, gender identity, religion, and otherwise), which shows just how false the idea is that only wealthy white people care about the quality of their food and their health. These claims are also downright insulting. Ultimately we are better served by understanding that the industrial food system affects all of us in profound ways. And while it's important to see and acknowledge how the industry preys on specific vulnerable groups, it's also safe to say that the industry is preying on all of us.

That said, we must also push beyond the simplistic citing of statistics to examine the power dynamics at work, and these do often operate along racial, class, and gender lines. This is not the end of the conversation but where fruitful discussion begins.

Issues of class, race, and gender are at the crux of our political food economy, and obtaining true food autonomy will depend on our recognition of all the ways the powerful industrial food system has manipulated us. It has completely upended American food culture to the point where simple, unadulterated foods are un-American or elitist and foods that are processed and full of environmental chemicals are true American fare.

The industry has also capitalized on one of our other shared American problems: lack of time, overwork, and the need for convenience. Perhaps the most telling common denominator, no matter the segment of the population the industries are appealing to, is convenience. Ease of consumption is directly related to the fact that it is no longer possible for middle- and working-class families to survive on a single income. In the past a family could afford for one parent to stay home and provide all the child care and housework necessary for raising a family. Fifty years ago the biggest employer in the United States was General Motors, where workers made the modern equivalent of $35 per hour. Now the biggest employer is Walmart, where workers make less than $9 per hour.[19] Indeed, Americans are working more than ever, with less to show for it. As Elizabeth Warren and Amelia Warren Tyagi document in their book *The Two-Income Trap*, middle-class families are now dependent on two incomes to buy homes, and while couples earn more money than their single-income counterparts a generation ago, they have less to spend.

The combination of the increasing time crunch on Americans with less money to spend on better quality foods is the perfect recipe for the food industry to capitalize with its convenient and cheap foods. And we can't address the issues of time and money without addressing the gendered nature of cooking, food shopping, and other household work. As of 2013, women are not only more likely to be the primary caregivers in a family; increasingly, they are primary breadwinners too. Four in ten American households with children younger than eighteen now include a mother who is either the sole or primary earner for her family.[20]

The related issues of time, housework, cooking, work outside the home, and gender roles in our society are critical to understanding how we might begin to end our national health crises. While others have argued that the industrial food system capitalized on the feminist movement by helping to get women out of the kitchen with its convenience foods, the takeover of American food and food preparation was part of the larger phenomenon in which industry sought a certain measure of control over what women bought, how they shopped for food, what they fed their families, and how mealtimes were structured. After all, what women brought into the home and embraced as part of the family's routine spelled profits for the food industry. Americans, and women in particular, have become dependent on large corporations for basic survival, relying on them to simplify the daily tasks and functions that we should be able to attend to ourselves: feed our babies their first foods, feed ourselves, and feed our families. This cuts across all racial and class categories; the mother's role in the home is incredibly important to public health, yet no one really addresses it. Women make up 90 percent of all caretakers in the

United States, so no matter your race or your social status, what your mother has access to—time, money, or knowledge of cooking and health—is going to directly affect your long-term health and overall quality of life.

Handmaidens of Industry

Women, the Home, and Unwaged Work

The basis for a healthy population begins in the home. The deterioration of our national health is a reflection of what has happened in the home, and particularly our kitchens, over the past several generations. We have undervalued the work done in the home and in particular "women's work."

With so much of our long-term health tied up in our crucial first foods and what we eat as children, we must acknowledge the political and public health functions of the home and motherhood. Yet the public health aspect of the home has been completely left out of the conversation about food, health, and politics—and the only way to change this is to assign real value to the work done in the home. When it comes to public health measures, breastfeeding, in addition to cooking and preparing healthy whole foods, is one of the most significant activities.

As we have seen, what we are exposed to in the womb and what our diets consist of as babies and children sets the stage for our future health. If you miss out on the transfer of protective bacteria from your mother during birth and breast-feeding, your immune system is compromised from the start and your lifelong health will likely suffer. We also know that what we feed children and babies—including breast milk—has a long-lasting influence on food preferences and food choices throughout life—babies who are not exposed to the diversity of flavors found in breast milk are less likely to eat a more healthful diet later on. And babies who are breast-fed are also significantly less likely to become obese than formula-fed infants; thus breast-feeding protects them from many diet-related diseases such as diabetes, heart disease, and some cancers.

While the decision to breast-feed or not has been framed as a personal choice, most women have little choice—they simply can't afford to. Only 12 percent of female workers and 5 percent of low-wage female workers have access to paid leave, which means that if they stay home with a newborn, most women will lose income they need. And the biggest determinant of who starts and continues breast-feeding is socioeconomic status. Seventy-four percent of children in families with incomes greater than 185 percent of the federal poverty line are breast-fed, compared with only 57 percent of children in families with incomes at or below the poverty line. Yet the vast majority of women want to breast-feed: one study found that 85 percent of women wanted to breast-feed exclusively for three months or more, but only 32 percent achieved that goal.

The United States does not require employers to provide parents with adequate leave time, yet other nations manage to—Bulgaria, Hungary, Japan, Lithuania, Austria, Czech Republic, Latvia, Norway, and Slovakia offer working parents more than *a year's worth* of paid leave.[1] The failure of the United States to do so is a prime example of how politicians on the left and the right pay only lip service to the importance of motherhood and families.

If mothers and fathers didn't have to decide between being able to pay their bills or staying home with newborns, how different would our country look? Not coincidentally, the United States is below average on most measures of health—from obesity to infant mortality—when compared to other rich nations.

Our industrial food supply compounds the personal and societal costs of the failure to provide broad paid parental leave benefits. As parents ponder what to feed their growing babies and children, their decisions are significantly influenced by an industry that has invented the billion-dollar markets of "baby food" and "kid food." The convenience and low cost of these foods are also significant lures: if parents are not in the home or do not have enough time away from work outside the home, they simply cannot do the shopping, cooking, and cleaning required to cook healthy whole foods meals from scratch.

Even progressives and "food movement" leaders think the way to solve our food-related health crises is to call on people to buy organic, cook more, and sit down every night for family dinners. In 2013 Michael Pollan, the food movement guru and all-around spokesman for everything food related, wrote an entire book titled *Cooked*, an homage to the importance of cooking. Throughout, he implores Americans to cook more, rightly pointing to the decline in cooking in the home and the rise in chronic diseases and poor health. The trouble is that he never details just how Americans, the majority of whom are working longer hours than ever and earning lower wages in general, might pull this off. The idea that we need to cook at home with fresh ingredients from scratch sounds simple but is actually exceedingly difficult for many Americans. What's more, Pollan emphasizes the importance of "slow foods," and he goes on lengthy explorations of the time-consuming processes of cheese making, bread baking, and fermenting his own vegetables. For Americans who are cooking hardly anything at all, the idea that they might take up the long processes of sourdough bread baking or sauerkraut fermenting is not realistic.*

Indeed, a 2010 report found that of thirty-four nationalities considered, Americans spend the least amount of time cooking, just thirty minutes a day.[2] This allows barely enough time to roast simple vegetables or bake some fish, never mind bake your own bread—an hours' long process. Pollan argues that with our loss of cooking skills we have lost something important about our humanity—and I agree. And his calls to get back in the kitchen and cook more are obviously imperative—but if we are serious about the importance of doing this, we have to address the time constraints faced by working Americans, the economics of being able to afford fresh whole foods grown without pesticides or other chemicals, the loss of cooking skills among recent generations of Americans, and the gender politics that are implicit in calls to get back into the kitchen and cook.

Pollan briefly addresses this last issue when he writes early in the book, "Now, for a man to criticize these developments [eating and using processed foods in place of home-cooked foods] will perhaps rankle some readers."[3] But Pollan doesn't delve much into

---

*After he read my May 10, 2013, op-ed piece in *The New York Times,* "Pay People to Cook at Home," Pollan tweeted that the solutions I offered were "not realistic" without offering an alternative.

the gender politics of cooking and the home in his nearly five-hundred-page book, but this is key to understanding what the industrial food system has done to our food, our health, and our bodies. The corporations now feeding us have undermined the biological process of breast-feeding through their promotion of infant formula, and they have drastically changed the types of foods that babies and children eat by advertising "white," or bland, foods—chicken tenders, pizza, and mac and cheese—thereby limiting variety and nutrition. In doing so, the food industry has stripped us of the important power and sense of responsibility and accomplishment that once came with the satisfaction of providing for ourselves and our families.

From a survival perspective, the only way we can stop cooking or breast-feeding is if someone else is doing it for us. Because corporations are now doing the cooking for the most part, we no longer have to cook at all. Pollan argues that one of the biggest reasons corporations were able to make their way into the kitchens and dining tables of America "is because home cooking had for so long been denigrated as 'women's work' and therefore not important enough for men and boys to learn." He adds that it's hard to say what came first: "Was home cooking denigrated because the work was mostly done by women, or did women get stuck doing most of the cooking because our culture denigrated the work?" But Pollan gets a key historical fact wrong here: Americans didn't always denigrate cooking, and before the twentieth century, American women were heralded for their domestic skills, especially their cooking.

In fact, the middle-class housewife of the nineteenth century, while not considered equal to her husband, was held in high regard.

By the 1850s, "the home" was a concept embedded in the national culture. What scholars often refer to as "the cult of domesticity" gave women a previously unknown political and social power. Men at the time (1820–60) respected the cult of domesticity, and the idea of home moved beyond the "women's sphere" and onto the national agenda, where it became "a touchstone of values for reforming the entire society," as the historian Glenna Matthews writes. One of the most fundamental aspects of homemaking, of course, was the housewife's ability to cook, bake, and prepare meals for her family. Matthews refers to the mid-nineteenth century as "the golden age of domesticity," and by that she means that middle-class women organized and influenced the political sphere as never before: "Public and private values were genuinely intermingled rather than being dichotomized."[4] Although women still could not vote and were far from achieving any semblance of equality, housewives became politically active, and men respected their opinions and valued their work. By this time, just as many girls as boys were receiving an education, and women's literacy enabled them to read more about domesticity and the home, topics that occupied even the most popular male writers of the time, including Ralph Waldo Emerson and Nathaniel Hawthorne. In fact, both men and women read articles and books about domesticity, the home, and its role in a democratic society.

But a profound shift occurred in the twentieth century when the home came to be seen as solely the concern of women, and domesticity was no longer deemed important by the political world. This shift coincides with the entrance of various industries to the home, in particular, the influence of the industrial food system and its products.

The commodification of the home took hold in the late 1800s as populations moved from a life based on farming and homesteading to life in urban areas. In 1820 only 7 percent of Americans lived in cities but 20 percent did only forty years later.[5] With industry producing many home goods and food products, women were able to purchase them rather than produce them. That's not to say that certain technological advancements along the way were not liberating to women; clearly, washing machines freed up an enormous amount of time for women. But unlike washing clothes, which is more like drudgery, cooking or baking from scratch is a skill—some would say an art—and a rewarding one at that. "If one examines the tasks of the housewife, one finds some, like laundry, that are akin to the traditionally male job of ditch-digging, and some, like cookery, that are more akin to cabinet-making," Matthews writes.[6]

Thus the entrance of industrial food products to the home marked a major shift in the role of women in the home, one that continued in the decades that followed as more and more families moved to urban centers. The 1920s saw the biggest increase in processed foods in the American kitchen, in large part because of a new technique for freezing foods that was patented by Clarence Birdseye. By 1934, U.S. food producers were processing thirty-nine million pounds of frozen foods annually, and freezing and canning created enormous changes in cooking and the American diet. Birdseye was among a handful of brand-name food companies whose products were becoming familiar in the American kitchen, thanks to the emergence of a national market and innovations in packaging.[7]

Historians situate the decline of American cooking even

earlier, though, dating it to the introduction of chemical leavenings in baking and artificial ingredients creeping into recipes as early as 1869. Cookbooks of the time began reflecting these changes, and most cookbook writers were now "handmaidens of industry, wittingly or no," as John and Karen Hess write in *The Taste of America*.[8] Matthews defines two distinct periods of industrialization's influence on American cooking. In the first period (the second quarter of the nineteenth century) cooks had access to more and a greater variety of ingredients, and the housewife also enjoyed more time to spend cooking "without any noticeable de-skilling of the cooking process." Middle class Americans were also developing "relatively sophisticated palates and had an appetite for baked goods made with sumptuous ingredients," she writes. But during the Gilded Age (1870s to 1900) "industrialization began to have the opposite effect, and a de-skilling process began, along with the concomitant deterioration of the cuisine. This process would accelerate in the twentieth century."[9]

By the 1920s the American housewife was portrayed as an agent of consumption rather than production. "Beginning in the 1920s, a new image of the American housewife took shape, an image suitable for a new age of material invention and consumption. The advertising industry, the manufacturers of household goods, the food companies, the women's magazines, and the schools all shared in the task of creating a woman who could discriminate among canned soups but who wouldn't ask too many questions about the ingredients," writes the author and food historian Laura Shapiro. Housewives were expected to trust the food manufacturers' assertions that their products were safe, healthy, and better than home-cooked versions. "Intelligent buying—meaning, of course, the homemaker's willingness to believe what a manufacturer chose to tell her—easily became a more important domestic function than intelligent cooking and cleaning," Shapiro continues.[10] The housewife's duties therefore went from being productive and skilled, as well as valued by society as a whole, to consumptive and rote, with little skill involved—essentially dismissed by the culture as unimportant and trivial. This was a direct result of the changes to society that came with the Gilded Age and the emergence of the industrial food system and its pervasiveness in the home, which rendered the ideology of domesticity obsolete. And while waged labor was becoming the norm in our industrialized society, women's work in the home never became waged labor.

But by the early and mid-twentieth century, unwaged women's work in the home was well established as a critical part of American capitalist economics. Acknowledging the importance of the role of the housewife in the functioning of the economy, one of the most prominent economists of the time, John Kenneth Galbraith, writes, "The conversion of women into a crypto-servant class was an economic accomplishment of the first importance. Menially employed servants were available only to a minority of the preindustrial population; the servant-wife is available, democratically, to almost the entire present male population." He goes on to say that if these servant-wives were paid, they would be the largest category of the workforce. When he wrote *Economics and the Public Purpose* in 1973, the average housewife did about $257 worth of work a week, or $13,364 a year (or $68,248 in 2011). A 2011 survey estimated that stay-at-home moms should be earning $115,000 a year for their work.[11]

UNPAID "WOMEN'S WORK"

According to the survey that found stay-at-home moms should be earning $115,000 a year for their work, the average mother works 97 hours a week in the home, including 13.2 hours as a day-care teacher, 3.9 hours as household CEO, 7.6 hours as a psychologist, 14.1 hours as a chef, 15.4 as a housekeeper, 6.6 hours doing laundry, 9.5 hours as a PC- or-Mac operator, 10.7 hours as a facilities manager, 7.8 hours as a janitor, and 7.8 hours driving the family car.[12]

Why was it that when all other forms of labor changed over to wage labor with industrialization, labor in the home remained unpaid? Selma James, cofounder of the Wages for Housework Campaign and a long-time activist, argues that once the factory became the center of production, "women, children, and the aged lost the relative power that derived from the family's dependence on their labor, *which had been seen to be social and necessary.*"[13] Thus anything that could not garner a wage was deemed unimportant, and because productive work in the home was in rapid decline, the best way for the home to serve capital was for it to become primarily a place of consumption.

At the same time, the food industry was determined to provide everything a woman once made herself in the home—its goal was to produce everything ready-made or at least partially made. Some products to which this idea has given birth are ridiculous: precut apple slices in individually wrapped plastic containers, frozen peanut-butter-and-jelly sandwiches, and Kraft's avocado-free guacamole, just to name a few. Americans are so far past the era of scratch cooking or baking that simply putting together your own peanut-butter-and-jelly sandwich, which is composed of three packaged food products in the first place, is too much trouble for many. But this is all part of the consistent devaluing of women's work in the home, which has allowed us as a society to ignore and make invisible all the work women do in the home, even though our entire economy is utterly dependent on it.

The de-skilling and devaluing of the role of the housewife also had a depressing effect on the women who did this work. It is what Betty Friedan wrote about in *The Feminine Mystique*: "The problem that has no name." Women became chauffeurs and shoppers rather than producers, cooks, and bakers—they were no longer engaged in any skilled work and were often isolated from family and community in suburban homes for long periods of time.[14] The 1950s saw the emergence of the "domestic goddess"—advertisements, magazines, and advice columns portrayed women as handmaidens to industry as they used the recipes that called for industrial products and stocked their pantries with processed food products. The domestic goddess was indeed an invention of the various industries—to a large degree the food industry—that sought to take over the American home.

A particular kind of gender oppression resulted from the industrial food system's infiltration of the home. This oppression placed the woman in charge of all household tasks and cooking, yet these tasks, which had been skilled work for the housewife of the 1800s, had become drudgery. Recipe books at the time reflect this shift—they call for cake mixes, canned foods, and soup mixes. The housewife's daily concerns were seen as trite and unimportant compared to her husband's day at work. It's no wonder, then, that women of the 1950s and '60s, referred to themselves as "just a housewife," as my maternal grandmother, who is now ninety, told me during a recent visit. It is also no wonder that many women of this era felt a certain emptiness or lack of purpose with this form of housework. Gone were the days of satisfying work that the women of the 1800s enjoyed in the home, which also gave them a certain status.

Also gone were the days of women, men, and children working side by side in the home, engaging with each other in a common purpose. It's worth remembering that although the sheer amount of work a home required before industrialization was tremendous, the idea that women were responsible for all the household duties is a distinctly modern one. In the preindustrial era the burden of housework fell on the shoulders of every inhabitant of the house: men, women, and children all devoted their energies and time to work in the home and on the farms, because most Americans were involved with at least some of their own food production. Although historians disagree about the relative power women held in colonial times, they were certainly not isolated and alone in the home all day and responsible for all the housework. The author and professor Maxine Margolis argues that there were no clear distinctions between what was considered women's work or men's work and no clear separation between home life and the world of work. "Male and female spheres were contiguous and often overlapped, and the demands of the domestic economy ensured that neither sex was excluded from productive labor. Fathers, moreover, took an active role in child rearing because they worked near the household," Margolis writes.[15]

Although men and women were certainly not seen as equals

during this time, women did have a voice in the distribution of economic resources, and they were valued as productive members of society.[16] With industrialization men left home for the factories and children left home for school, placing the entire burden of housework on women (though in some cases women and children went off to the factories too, and they were exploited to a much greater degree than the men were). Women, though, whether they went into the factories or not, were expected to be the sole household caretakers, doing almost all the work. This expectation persisted and intensified in the decades that followed and by the 1950s and '60s, many housewives were alone in the home all day, isolated from community in sprawling suburbs, and left with the more menial and rote household tasks.

As such, the argument that Pollan and others have put forward—that, by providing women with convenience foods, the industrial food system capitalized on the desires of women to get out of the home and into the workplace—is not correctly formulated or historically accurate. The corporate takeover of the American food supply actually began much earlier and was part of the larger campaign in which the food industry stepped in for the role of women in the home by offering goods like canned foods, frozen foods, precooked or partially cooked meal options, cake mixes, and store-bought bread, as well as providing infant formula in place of breast milk. These changes began long before second wave feminism of the 1960s and 1970s, which prioritized getting women out of the home and into the workforce. And by that point, who could blame women for wanting to get out of the empty houses where they were confined for most of the day?

Especially once children went off to school, women were often alone in the home without any satisfying or productive work to do. But had the role of the home maintained some of the prestige it held in the 1800s, and had women maintained some of the craft skills of cooking and other household tasks, along with the social and cultural capital it commanded, perhaps the feminism of the 1960s and '70s would not have been so focused on getting women out of the home and into the workforce—perhaps that feminist movement could have also placed value on the *work done in the home* and considered just how much it was worth to society.

Second wave feminism, therefore, made a tactical error by emphasizing that the key to women's liberation was getting out of the home and into the workforce. This prioritized the middle class, in particular white women, because getting predominately white middle-class women out of the home meant that women of color and women of lower economic status were coming into the home in greater numbers to do the work of cooking, cleaning, and child care. This is hardly liberation for all women. Even in the late nineteenth century this was becoming a major impediment to liberation. "Women can never gain their own liberation from stereotypes of gender at the expense of other women of a lower economic class or another race whom they exploit by paying them low wages to do sex-stereotyped work," the professor and author Dolores Hayden writes of that period. "Black women and white women, Yankee women and immigrant women, housewives and servants, had to break out of woman's sphere together, or else not at all."[17] Furthermore, when women went off to work, they received second-class treatment, took more menial jobs, and received less pay than men—trends that continue to this day. Obtaining second-class status in the workforce while still being expected to do the household work, or the notorious "second shift," is hardly an improvement. But the goal of more egalitarian households could have taken shape had corporate interests not invaded the American home with such swiftness and force and if the feminist movement had acknowledged the importance of work in the home.

Instead household work was further devalued, and the burden was placed largely on the shoulders of poorly paid women of lower socioeconomic status. In many ways this legacy persists from the preindustrial era to today. The high regard and respect afforded housewives at the height of the cult of domesticity was mostly reserved for middle-class and upper-middle-class white women. Particularly in the southern United States, black women did much of the work in the kitchens but were not given credit for it. Black women were cooking, cleaning, and taking care of the children and were responsible for most other household work. In many instances, servants were serving as wet nurses, breast-feeding the children in middle-class white homes. Cookbooks provide a window into the domestic culture of the antebellum years. Their authors were largely white women, who often incorporated the cooking and recipes of black women who worked in their homes. The food journalist Toni Tippi-Martin, author of *The Jemima Code: Two Centuries of African American Cookbooks*, found that of the 100,000 recipe collections produced in the United States since the late 1700s, only 300 were written by or attributed to African Americans. Many of the most popular cookbooks became sensations. White women got the credit. Their black cooks remained invisible.[18]

Even today a walk through Prospect Park, Brooklyn, reveals this reality. Nearly everywhere you look, women of color are pushing strollers with white babies or playing with white children. Who's caring for these women's children? And what kind of effect does this have on these women's home lives? One reason that wealthier couples achieve greater gender equality in the home is because they rely on the labor of poorly paid women to do the gendered work of child care and housework.[19] But one major difference in today's version of this dynamic is that no one is really cooking in the home at all—these caretakers are feeding babies and children the same low-quality baby and kid foods that their busy parents do. And caretakers are certainly not breast-feeding other women's babies, which would now be considered strange or unsanitary—but these bottle-fed infants would be far better off if they were receiving breast milk of any kind rather than none at all. It's true that many mothers pump breast milk to give to their babies while they are away at work, and this is laudable. But wouldn't it be better if they had the time and resources that allowed them to breast-feed infants for at least six months at home? These mothers also tend to be women who can afford a full-time nanny and have the resources and encouragement in the workplace to pump breast milk for their babies. Pumping breast milk takes time, and many employers frown upon women taking extra-long breaks to do it. The vast majority of working women, however, are not pumping and are instead relying on formula. Remember: the percentage of women who exclusively breast-feed at six months is an abysmal 18.8 percent, even though many leading public health authorities, including the American Academy of Pediatrics, strongly recommend breast-feeding.

It's hard to overstate just how much our health depends on our diet, particularly our first foods. By removing women from the home and by replacing breast milk and home cooking with their industrial counterparts, we have significantly altered the culture around the home and motherhood as well as the status of our health. So much is threatening our health that we must address the economic factors that continue to shape this new reality. So long as working- and middle-class families need two incomes to survive, parents will do little cooking in the home, and not many women have the luxury of staying home to breast-feed.

Yet a certain mythology persists that women can do it all. Perhaps most famously, Sheryl Sandberg's bestseller *Lean In* tells women that they can succeed in corporate America while also fulfilling the roles of wife and mother. For most women, especially those not in the stratum occupied by Sandberg, this is unrealistic. Indeed, for every woman as successful at paid work as Sandberg, four women are holding down traditional women's jobs, such as secretary and elementary school teacher. To this day the most common job for women is secretary, or as it is called now, administrative assistant, whose duties remain unchanged since 1950. Ninety-six percent of administrative assistants are women, and their average salary is $30,000. Fewer than 20 percent of chief executives are women.[20]

And as women "lean in" to their jobs outside the home, certain aspects of mothering are clearly lost. Men could step in to do the other household work, but not breast-feeding. And so a certain educated class of working mothers has rejected breast-feeding—women who believe that breast-feeding represents an almost antifeminist position, tying women to their babies and to the home and undermining the feminist revolution of the 1960s and 1970s. It's an interesting phenomenon, almost a backlash to a backlash of sorts. Where many "earth mothers" were coming back to breast-feeding and embracing a more natural form of child rearing in the 1970s after the industrialization of these processes in the 1950s, some women of the later 2000s and 2010s are once again rejecting a more natural approach. The 2015 book *Lactivism: How Feminists and Fundamentalists, Hippies and Yuppies, and Physicians and Politicians Made Breastfeeding Big Business and Bad Policy* represents much of this thinking. Demonizing breast-feeding as an antifeminist activity is extremely odd thinking.

Selma James, who began the Wages for Housework Campaign in the 1950s, told me in a 2015 interview that she has encountered this attitude many times, particularly from professional women. "It has become common that breast-feeding, which is women's first contribution to society, the biological contribution, and the caring work that follows, is demeaned in favor of the work women do outside of the home," she said. "What is the elevation of these professions that has the audacity to demean breast-feeding and prioritize itself over breast-feeding? There is something radically wrong with where feminism has gone."

This brings us back to the question that Pollan poses in *Cooked*: "Was home cooking denigrated because the work was mostly done by women, or did women get stuck doing most of the cooking because our culture denigrated the work?" The framing of this question elides another key issue: the fact that food is so deeply connected to women and especially the role of mother. "An infant's first relationship with another person is based on the experience of one body feeding another body," writes the historian Janet Flammang. And as much as we do not like to make essentialist arguments about the roles of women and men, there is no denying the biology of pregnancy, childbirth, and breast-feeding. "Food is profoundly associated with women because our early memories include first food coming from women's bodies." Perhaps one of the pitfalls of feminism is found here: the desire to remove oppressive constructs around the roles of both women and men has undermined a certain power inherent in the role of mother and woman. "Food preparation is a labor-intensive activity. Bodies have to do it. The question is, which bodies?" Flammang asks.[21]

The industrial food system, though, in its efforts to overtake the American kitchen and in essence to replace the role of mother with infant formula and prepackaged meals, in many ways disembodied food production entirely by making it mechanistic, neat, clean, easy, and completely devoid of any requirement of the senses. This is the antithesis of breast-feeding, which is the most embodied form of feeding possible—the mother literally dissolves her own body to make food for her baby. Breast milk is also the most nourishing and protective food of all—and that's not just some romantic notion. Researchers in the fields of human nutrition, breast milk, and the microbiota know this to be true. As Bruce German, the professor of food science and technology at the University of California, Davis, told me, "In the broader sense, science as an enterprise does not know how to nourish people at any age."[22]

In many respects the entrance of science into the home in the mid- to late 1800s not only undermined the vast importance of breast-feeding but marked the beginning of the end for the craft tradition of cooking by introducing domestic sciences, or home economics. Domestic science and the home economist placed an emphasis on efficiency and devising a mathematically precise approach to housework, cooking, and child rearing. The home economists of the first decade of the twentieth century sought to strip the work of any sentimentality or sensual enjoyment. "A favorite device was to imply that there was something suspiciously sensual—and not at all scientific!—about a too-enthusiastic enjoyment of food," Matthews writes. The birth of home economics was part of a larger pattern of professionalism that came about in the late nineteenth century, and its foundation was the idea that the more rational, mechanized, and therefore masculine the work was, the more value and importance it had. The home economics phenomenon also had the effect of separating the mind-body duality that especially linked the process of food production. "The devaluing of food and the body in favor of higher-order activities stubbornly persists in dualistic thinking," Flammang writes.[23]

The separation of women's relationships from the products of their labor, from cooking, preparing, or growing and raising food, to the process of breast-feeding, seems to be correlated with women's ability to maintain a certain level of visibility and respect within the culture. By looking at nonindustrial societies, including the Iroquois of early America and the !Kung of contemporary southwest Africa, anthropologists have found that "the status of women as laborers is less a function of the subsistence value they create than it is of their ability to control the products of that work. And that ability appears to be closely linked to gender, especially as expressed in the operation of kinship groups. So long as the kin group is organized, both literally and symbolically, in a way that acknowledges women's claims to the products of their labor, women appear to be able to maintain visibility as workers," Boydston writes.[24] This holds insight for what happened to women's labor in the home with the introduction of industrial food products. Women could no longer claim the products of their labor when corporations produced those products en masse. And since the food companies now controlled the products of women's labor, they stripped women of their cultural status and visibility that came from doing productive and valuable work in the home.

This helps explain why such a vehement rejection of the "woman as housewife" occurred in the 1960s and '70s, since it was during this period that the separation of home and work peaked, as did the devaluation of women's work in the home. And now the role of the housewife, or the person who does the cooking and shopping, has been so undervalued for so many years that it is basically obsolete among most Americans. That is, no one cooks, and if they do, it is a much-abridged version of cooking, and for the most part no one even notices. Wealthier families simply pay people to cook for them, either in the home or in high-end restaurants, while poorer families pay the food industry to cook for them with its highly processed convenience food and fast-food meals. In both cases *someone* is being paid. But no one is paying the outliers, the women and men who do still cook at home from scratch with whole food ingredients. At this point in our history—after hundreds of years of consistently devaluing household work, which is chiefly done by women, especially cooking—how can we remedy the problem of unpaid labor in the home? And why do we continue to accept that this labor is unpaid?

To start, paid parental leave must be mandatory. But even the twelve weeks of paid family leave that Bernie Sanders called for are not enough. Three months at home with a newborn is superior to the six weeks afforded by some employers, but we will need something far more radical to make a significant impact on working parents in this country and to reverse our public health crises, which are now of epic proportions. Leading health agencies all agree that lifelong health requires at least six months of breastfeeding. Thus a minimum of six months of paid parental leave is an obvious first step.

Feminists like Selma James began arguing in the 1950s and 1960s that without the role of mother, wife, and caretaker, capitalism would cease to function. This is clearly what the economist John Kenneth Galbraith meant when he said that the capitalist economy requires a "crypto-servant class"—wives. He wrote that the crypto-servant class made possible the indefinite expansion of consumption. "One of the singular achievements of the planning system has been in winning acceptance by women of such a crypto-servant role . . . And in excluding such labor from economic calculations and burying the separate personality of the woman in the concept of household, where her sacrifice of individual choice goes unnoticed."[25] How much longer will we accept the narrative of choice when it comes to all the work of motherhood, breastfeeding, and cooking for children and our families? Whether we do it ourselves for no pay or pay other women a low wage to do it for us, we are aiding a food industry that cares only about its profits. More than ever, we need innovative and radical ideas that challenge our ideas about gender, food, cooking, and our health. We must first acknowledge that the work we do—or don't do—in the home is at the very foundation of public health and the health of our society as a whole.

We Can't Eat Our Way Out of This

Food Movement Failures and Real Change

It's a common sight: a baby drinking out of a plastic bottle, a toddler drinking from a plastic sippy cup or a fruit squeeze pack—but most parents think nothing of it. That is especially true now that these bottles, cups, and packaging are labeled BPA-FREE, after advocacy groups fought for years to rid baby and kids' products of the notorious endocrine-disrupting chemical (EDC) bisphenol A, which is ubiquitous in plastics, metal can liners, and a range of food packaging. You'll see the label displayed prominently on thousands of products on grocery store shelves. In the early 2000s "mom groups" were especially vocal about keeping the chemical out of children's foods because of the damage it appears to do to the developing brain and reproductive and metabolic systems. And in this case, advocates achieved a seeming victory when the Food and Drug Administration announced in 2012 that industry could no longer use the chemical in baby bottles and children's sippy cups. But then the industry replaced the chemical with bisphenol S—a related chemical compound that not surprisingly has the same kind of deleterious effects in the bodies of children and adults alike.

"It's an empty victory. I feel bad saying that because there was really strong work on the part of advocacy groups that fought to get rid of BPA, but they would be horrified to know that their actions replaced one estrogenic chemical with another, and yet it became a marketing tool," said Laura Vandenberg, an EDC researcher who teaches at the University of Massachusetts. Since then, "BPA-free" has been used as a marketing tactic, while the food and chemical industries sells us another environmental chemical as progress. This is exactly what is wrong with allowing the free market to make what are essentially public health decisions. Change within this paradigm is limited to change that maintains profits for corporations—we didn't rid our foods of a toxic chemical, we just replaced it with a different one.

Indeed, the FDA said that the ban did not mean the agency had changed its determination about the safety of BPA—it said it was safe in 2008—it simply codified what the industry was already doing in response to consumer demand.[1] This is also why we can't rely on consumer demand and individual buying decisions to significantly improve our food supply or make it any safer. The case also highlights just how illusory the idea of choice is within the framework of the industrial food system. We certainly may be choosing what products to eat, but by and large we don't have much of a say in what, exactly, goes into those products, nor are we properly informed about what they contain. The food industry's claims of personal responsibility and freedom to choose are clearly false.

Yet there is another false narrative that may be even more problematic because it comes from the very people who rail against the food industry. Food movement leaders and activists often argue for buying organic and cooking more at home (they less often advise avoiding foods with environmental chemicals like BPA), but these solutions are overly simplistic, near-impossible for the vast majority of Americans, and do not result in real change. What's more, the food movement's emphasis on providing access to and education about healthy whole foods is deeply problematic in the face of economic hardship, lower wages, overwork, and a complete lack of time. These popular food movement refrains just represent narrative shifts, not actual strategies for improving our food supply for the majority of Americans.

When food movement leaders say the solutions are to eat whole foods and buy organic, they leave out the crucial fact that we need to collectively reject the production of poor-quality processed foods and stop the production of dangerous pesticides and other environmental chemicals that contaminate many foods. Critics do not often articulate this omission, but it is largely why the movement is perceived as elitist, and rightly so. If the food movement's solutions are market based and predicated on spending more for safer and healthier food, they ignore how impossible these solutions are for most Americans. In fact, this approach serves the agendas of Big Food and Big Ag quite well. The food movement's insistence on organics for those who can afford them has allowed the industry to profit on multiple fronts by creating a booming organic market while also maintaining the market for its less expensive conventional products. A far more powerful approach would be to demand an end to the thousands of dangerous food products on grocery store shelves. And a particular focus should be ridding the food supply of the most dangerous additives, such as the various pesticides, herbicides, fungicides, and food-packaging chemicals that I have discussed throughout this book. Instead, the food movement has allowed these products and additives to exist along-side a cleaner and safer food supply for the privileged few.

Food movement leaders also emphasize the importance of home cooking and cooking whole foods from scratch. Yet many fail to mention that the majority of Americans do not have the time, money, or resources to cook meals from whole foods at home. And when these leaders do acknowledge that lack of time to cook is a problem, they usually address it through providing better ways to cook healthy foods quickly. Responding to the time crunch with easily marketed advice for "quick and easy" ways to prepare healthy meals is not the solution—instead, we must reorganize our lives so that we have more time for crucial life-sustaining skills like cooking healthy meals and nourishing newborns and children properly.

Meal kit companies like Blue Apron and cookbook authors sell the idea of easy, fast, healthy cooking. But these tactics en-gage the capitalist ethos that says every market-based problem has a market-based solution. And it further sends the message that you should cram healthy cooking into your already busy daily life—rather than change your day-to-day lived reality. But the truth is that adding even an additional thirty minutes of cooking to the day is impossible for many Americans.

This is especially true when you consider that nearly half of the entire U.S. population is struggling to survive on poverty-level wages. More than 40 percent of all American workers make less than $15 an hour, and more than half of all women workers, African American workers, and Latino workers earn less than $15 an hour.[2] Forty million Americans rely on SNAP, or food stamps, to survive, and the National School Lunch Program feeds thirty-one million schoolchildren breakfast and lunch every day—that's nearly *half of all schoolchildren in the entire country.* These safety net programs are critical, but they do not solve the underlying problems. If nearly half of American workers make less than $15 an hour, how can we ask them to pay more for food and spend more time cooking? That's nearly 160 million Americans who don't earn a living wage. Women bear the brunt of this reality. Women across the board earn 77 cents for every dollar a white man earns, but black women earn only 64 cents, and Latino women earn only 55 cents.[3] At the same time, the majority of women are primary caregivers in the home. Telling the millions of women in this position to buy organic foods and spend more time in the kitchen cooking meals from scratch is not only impractical, it's insulting.

The welfare state is eroding, and the few existing safety net programs for women and children are inadequate at best. What's left is a shell of the Social Security Act of 1935, which as part of the New Deal implemented a program called Aid to Families with Dependent Children (AFDC) that gave cash grants to single parents so they could stay at home and care for their children. In the 1960s this program was expanded somewhat, but Republicans claimed it offered incentives for women to have children and discouraged women from joining the workforce. But as Barbara Ehrenreich made clear in her book *Nickel and Dimed*, the argument that welfare makes people lazy or provides incentives for women to have children has a fatal flaw: welfare simply does not provide enough money for one person to live on alone, never mind a family (the median welfare package in the United States is about $28,800 a year).[4]

AFDC was replaced by the more restrictive Temporary Assistance to Needy Families (TANF), which was part of the Personal Responsibility and Work Opportunity Act instituted under President Bill Clinton in 1996. There is that pesky "personal responsibility" concept again, and it did real damage by undermining a program that had long helped families in need, particularly those headed by single mothers. The program's budget was set at $16.5 billion per year and has not gone up since 1996. And when the Clinton administration revised the program, it became a block grant program, not an entitlement program, meaning that the money now went to the states to use at their discretion. Currently, less than 25 percent of the money set aside for TANF goes to needy individuals. In some states the money goes to "marriage promotion classes" and to "family planning counseling," which is meant to reduce abortion rates. A current effort would also turn SNAP into block grants, which would further undermine food security for the forty million Americans now relying on food stamps. School lunch programs are also potentially on the chopping block.[5]

One in three women in this country is living in poverty or on the brink of it, according to *The Shriver Report.* That's forty-two million women and their twenty-eight million children. The report also found that two-thirds of minimum-wage workers are women who get no paid sick days, and two-thirds of American women are either the primary or co-breadwinners of their families. When surveyed, 96 percent of single mothers said paid leave is the workplace reform that would help them the most, and 60 percent of low-income women said they believe that, even if they make all the right choices, "the economy doesn't work for someone like me."[6] Clearly, our society needs a major change. How can we undo cycles of poverty in this country in a way that would also address the alarming health disparities for low-income people? Income inequality and poor health are two of the biggest issues we will face as a nation in years to come, and we have to start thinking of big solutions to these seemingly impossible dilemmas. Part of the solution will reside in our ability to radicalize the existing food movement to agitate for real, substantive change—not the so-called change based on ideas like "voting with your fork."

The issues of poverty, public health, and nutrition are inextricably linked, and they must be addressed simultaneously if we hope to effectively address any of them. And the issues of work, housework, and gender roles are inextricable from the realities of poverty and poor nutrition—which means we will need multipronged plans of attack to address all these issues at once. First we will need to address income inequality through some type of universal basic income; we will need to figure out how to return the sources of our food supply to local and regional communities so that fresh, chemical-free produce, animal products, and other whole foods are available and affordable in all neighborhoods all the time; and we will need to teach people how to shop for, prepare, and cook healthy whole food meals. We will also need more worker-owned and -operated food cooperatives that make healthy whole foods available at prices lower than the supermarkets', and we need to finally address the economics of "women's work," which in large part entails the necessary and life-giving activities of shopping for food, cooking, and cleaning. So what will this look like? Beyond a push for a universal basic income (UBI), which has been written about widely, here are some other ideas that would complement a UBI.[7]

An example of a program that could begin to address the gender imbalances in care work, especially food shopping, cooking, and cleaning, is a program that would "Pay People to Cook at Home" (the title of my 2013 *New York Times* op-ed where I introduce the idea). The truth is that no real food movement is possible unless one or two people in a family do the work of meal planning and shopping and then preparing whole food meals from scratch.[8]

The kernel of this idea comes from the important and often overlooked Wages for Housework Campaign. Founded in London in 1972, its fundamental demand was to acknowledge the work that women do in the home and to demand wages and therefore financial independence for doing it. A groundbreaking text, *The Power of Women and the Subversion of the Community* (1971), by Selma James and Mariarosa Dalla Costa, outlines the centrality of unwaged work to the economy and society and the division between women and men. In the foreword, the group called the Power of Women Collective wrote that unwaged work in the home is the basis for the powerlessness of women in relation to both men and to capital. "Wages for that [house]work, which alone will make it possible for us to reject that work, must be our lever of power," the group wrote. "If our need for a wage and our need to break from our isolation have driven us to a *second* job outside the home, to work more at low pay, then our alternative to isolation and wagelessness must be *a social struggle for the wage*." In the introduction James argues that the family is the center of social production under capitalism (something the economist Kenneth Galbraith said as well). "If your production is vital for capitalism, refusing to produce, refusing to work, is a fundamental lever of social power," she writes, adding, "The commodity they [housewives] produce, unlike all other commodities, is unique to capitalism: the living human being—'the laborer himself.'" But while the housewife does the very work that allows capitalist society to run—by producing the laborers—"*the [housewives'] labor appears to be a personal service outside of capital*," James and Dalla Costa write.

Wages for Housework provides the crucial framework for questioning the very notion that second wave feminists put forth: that women will achieve liberation through work outside the home. The legacy of the tactical mistake that second wave feminists made—by fighting *only* for women to get out of the house and into the workforce—persists to this day, as the statistics about women working at poverty rates grimly illustrate. Most housewives in the industrial world now also work outside the home but for significantly less pay as second-class citizens. And with one in three women in America now living in poverty, it is plain to see that the push to get women into the workforce as a route to liberation undermined the well-being of mothers, children, and the family.[9] James and Dalla Costa write, "The challenge to the women's movement is to find modes of struggle which, while they liberate women from the home, at the same time avoid a double slavery and prevent another degree of capitalist control and regimentation. *This ultimately is the dividing line between reformism and revolutionary politics within the women's movement.*"[10] This assertion is just as applicable today as it was in 1972.

Ultimately, this gets at the idea of overwork for all. How is it that so many can work a full forty-hour workweek or more but still be in poverty? How can we live in a land of obscene excess and wealth while many people still do not have enough food to eat—and that is *enough of any kind of food*, never mind healthy whole foods. This epidemic of overwork is completely at odds with our health and with public health as a whole.

It is not only low-wage workers who are working more than ever; a 2013 survey of 483 executives, managers, and professionals found that most of them are working about seventy-two hours a week. If these people sleep about seven hours a night, they have only three hours each weekday for everything else.[11] No wonder diet-related diseases are on the rise across all population groups in the country.

Studies conducted by the Swedish Environmental Protection Agency found that a thirty-hour workweek actually increased productivity and improved all measures of quality of life, including overall life satisfaction and health, and helped to lower greenhouse gas emissions.[12] Mothers and families must have some relief from overwork so that our overall quality of life and public health can improve. This is fundamental to addressing our ability to overcome the crisis in food quality and health in this country, because without the time, none of the health-promoting work of breastfeeding, cooking, or caring for children is possible.

While U.S. workers are only guaranteed twelve weeks of *unpaid* parental leave (but only those covered by the Family and Medical Leave Act of 1993), several European countries, as well as

Japan, provide excellent models for reducing the pressures on working families. The Netherlands promotes a "1.5 jobs model," which allows and encourages both men and women who have young children to work 75 percent of their regular hours for full-time pay. Dutch collective agreements offer employees the choice of additional income, days off, or a leave period. Swedish parents can take leave for as long as fifteen months for each child at 80 percent of their pay, and they may do so at any time until the child is eight years old. In Norway parents can take forty-two weeks of leave at 100 percent of their wage, or fifty-two weeks at 80 percent. Norway also has what's known as a "father's month," or thirty days reserved for the father to take parental leave, which encourages fathers to be engaged in child care.[13] Sweden and Norway adopted a similar policy, and now more than 90 percent of men take parental leave.[14] Japan provides fifty-two weeks of paid leave for both mothers and fathers at about 60 percent of their salary.[15] And in the Netherlands, Norway, and Sweden, where these policies are in place, the obesity rate among adults hovers at about 12 percent; in Japan it is about 3.5 percent—significantly lower than the obesity rate of 34.9 percent for all U.S. adults.[16]

Both Spain and Trinidad and Tobago have introduced legislation to provide wages for housework, and time-use surveys are under way in many countries.[17] One article of Venezuela's constitution recognizes work in the home as an economic activity that entitles housewives to social security, because their work creates added value and produces wealth, thus contributing to social welfare. In recent years, members of the U.S. Congress have introduced two bills acknowledging that caregiving work should be compensated with wages.

RECENTLY PROPOSED LEGISLATION

The RISE Out of Poverty Act, introduced by Gwen Moore (D-WI), in 2011, 2013, and 2015, would make ending child poverty the primary goal of Temporary Assistance for Needy Families (TANF, the latest incarnation of welfare), raise benefits to meet families' basic needs, and get rid of some of the most punitive aspects of TANF.[18] The Women's Option to Raise Kids (WORK) Act, which was introduced in 2011 and 2013 by former representative Pete Stark (D-CA), recognized that raising children is vital work and would give mothers and other caregivers the option of staying at home full time to raise children younger than three without having to take another job under current welfare legislation.[19]

If we truly value domestic work and its contributions to our health, the economy, and public health as a whole, we need to implement social programs that prioritize the unpaid work of parents raising children to become healthy adults. The program that I am suggesting would include fathers and mothers who would like to opt out of the traditional workforce to take care of small children and household duties, cooking chief among them. This program would not be welfare; it would be available to all interested Americans regardless of income or wealth. Participants would attend classes in basic household skills with an emphasis on teaching healthy cooking from whole food ingredients. Completion of these courses would certify participants for pay for cooking and other household work.

Critics are quick to dismiss a program like this, saying that it will be rife with abuse. But this is the same argument that has been used for decades to condemn all social programs meant to provide basic services for those in need. And, frankly, wouldn't we rather deal with some level of abuse in a program that on balance could create enormous benefits to society? Critics also ask where the funding will come from, a valid question but one that ignores our bloated military budget, which in 2017 alone was increased by $80 billion, putting it at about $700 billion annually—three times what China spends and ten times what Russia spends.[20] And with the Trump administration, many experts predict another $250 billion increase over the next four years.[21] These are absurd numbers but ones that show what we value as a society.

Taxing the food and chemical industries, which make billions off our food consumption, could be another way to generate revenue for the program. We could create a corporate tax for the makers of poor quality foods, whether low in nutrients or high in sugar, that could fund a public program to encourage healthy eating and cooking in the home. The program would benefit the economy because obesity, the biggest driver of diet-related disease, costs the U.S. economy an estimated $210 billion a year.[22] Cardiovascular disease and stroke, cancer, and diabetes cost an additional $781 billion—all told, that's nearly $1 trillion a year.[23] And a report in the journal *Pediatrics* found that the United States would save $13 billion a year in medical costs if 90 percent of families breastfed their newborns for at least six months.[24] Why not invest some of that money in preventative measures? And as I have shown, prevention in the home—by providing parental leave so women can breast-feed and parents can cook whole foods—could be the biggest public health measure the United States has ever put into place.

Another idea to complement Pay People to Cook at Home by addressing the lack of fresh healthy foods in neighborhoods across the country—we could call it City of Farmers—is a federal farm program that would use vacant lots, public parks, and private yards for growing food in every neighborhood. The New York City Department of City Planning estimates that Brooklyn alone has 596 acres of vacant public land, and across America there are thirty-five million acres of lawn and thirty-six million acres for housing and feeding recreational horses.[25] The ultimate goal of this program would be self-sustaining farms in every neighborhood. There are many urban farm innovators at work in cities across America, but there is no united collective effort to create a web of farms across the country.

A federal urban farm program could result in paid positions on each farm, creating jobs for people in each community. The program could be set up like AmeriCorps, with college students and recent college graduates building and creating small-scale farms with community members—community service that would help offset some costs of college tuition or help repay student loans. Nutrition professionals and chefs could host nutrition, food preparation, and cooking workshops that highlight the foods coming off the farms. In some locations, there could be cooking and eating facilities for community members to host their own cooking demonstrations and feature their own recipes using food produced on the farm. Those interested could eventually move into permanent positions hosting the weekly workshops as paid employees. Eventually these farms could occupy every neighborhood, creating a web of farms across the country, not only making fresh foods available to all but adding vitally needed green space in cities for humans and wildlife alike.

A national food-growing program would be a way to solve the

problem of food deserts. Current programs are only temporary fixes: food banks and hunger relief organizations and school breakfast and lunch programs feed hungry people but do so mostly with the same industrial foods that are causing health problems. Food banks have to rely on what is donated, which is usually processed and packaged foods. The logistics of collecting and distributing perishables is also daunting for these organizations. A program like City of Farmers would help to rectify these problems and could also donate some of its food to hunger relief organizations until expansion of the farming program renders the relief organizations unnecessary.

Many community leaders and programs are already working toward this goal; connecting and unifying them could be powerful. Food Corps, which works to bring food and nutrition education to schools across the country, receives a portion of its funding from AmeriCorps and is a good model to build on. The individual efforts of visionaries like Will Allen in Milwaukee, Malik Yakini in Detroit, Ron Finely in Los Angeles, and Karen Washington in the Bronx, to name just a few, are vital to bringing food and resources to underserved areas and educating people about food and health.

One of the main problems with our current industrial food system is its centralization. When food products, or raw ingredients for food, like corn, soy, or wheat, are grown on massive industrial farms far removed from the communities they feed, we have no control or say in how these foods are produced. It also means that their nutritional content is compromised, and they are more likely to contain various preservatives and/or additives or to have been sprayed with harmful chemicals. Buying from local farmers also supports the local economy, rather than supporting multinational food corporations whose products now fill grocery store shelves. It's also a national security issue because the centralization of food makes it vulnerable to both deliberate and unintentional acts of contamination. Widespread food recalls and food-borne pathogens are difficult to trace and control when centralized warehouses are shipping foods from industrial farms all over the country and world. What's more, if an unforeseen event disrupted our food channels, most of us would be in a precarious position because grocery store shelves hold only about two to three days' worth of food. Ultimately, the more local your food, the better.

On this front, Will Allen's organization in Milwaukee provides the perfect model to replicate across the country. Allen believes that to revolutionize American farming, we will need to make small-scale farming economically viable, which will mean farming multiple crops intensively, even on small plots.[26] His organization, Growing Power, now has farms throughout Wisconsin and Illinois in both urban and rural areas. These farms grow dozens of vegetables, include an intensive fish-farming system known as aquaponics, and raise chickens, goats, bees, and worms, which are essential for creating fertile soil from compost. One of Allen's farms in the middle of Chicago has fifty goats, and his farm in Milwaukee has more than three hundred chickens. While this may seem to be at odds with our ideas about what a city is, recall that as late as 1880 Brooklyn and Queens were the largest producers of vegetables in the country, and farmers raised livestock in urban centers. Allen's vermiculture system (the worm farming) is a crucial way to reduce waste generated by our current food system. The United Nations recently estimated that nearly one-third of all food produced globally is wasted, at enormous economic and environmental costs.[27] Food sent to the landfill is a great contributor to global greenhouse emissions—it represents twice the amount of carbon emitted by the U.S. transportation sector alone.[28]

Allen's farms are examples of closed-loop systems; each farm incorporates crops of vegetables, animal husbandry, encouraging beneficial insects, and composting of its own waste. Growing Power also composts more than forty-four million pounds of local food and organic waste. In Chicago Allen even started his own composting facility, which composts five tons of waste a week from local restaurants, breweries, and organic wholesalers. And Allen's farms generate large amounts of food: their greenhouses produce more than one million pounds of fresh microgreens, fruits, and vegetables year-round.[29]

With many urban centers in decline, empty lots sitting unused, people looking for jobs, and public health in crisis, the solutions are right in front of us. City governments should be developing local farming and food policies. This is not a new concept. Biodynamic agriculture, also sometimes called permaculture, focuses on maximum yields in small spaces while also increasing biodiversity and increasing soil fertility. The ancient Chinese, Greeks, and Mayans used these techniques, which also have been used in Europe and West Africa since at least the eighteenth century. A related style of farming began on two acres near Paris in 1890 and became known as the French intensive method. This method was then introduced in California in the late 1960s and 1970s by Alan

Chadwick.* The point is that the models already exist; they just need to be replicated in cities across the country.

Pay People to Cook at Home and City of Farmers would not only be important models but important political perspectives for organizing and demanding change, much like James's Wages for Housework. Silvia Federici, another proponent of wages for housework and a contemporary and colleague of Selma James's, said we must think of wages for housework as a political perspective as well as a program proposal.[30] Thinking of wages for housework as a tool for deepening the conversation about the political structure of our economy, health, and the ability to provide healthy food for our families opens up a whole new discussion about food in this country. Suddenly the public health message changes from asking people to buy organic foods and shop at their local farmer's market to asking how our economy can better serve public health and the health of our families. These are two vastly different ways of thinking and participating in our political economy. Ultimately we cannot solve our environmental and public health crises without addressing the economic limitations that most Americans face today.

Another way to begin organizing around food immediately is by creating member-worker–owned food cooperatives. The Park Slope Food Coop, which I have mentioned throughout the book, is an important model for making healthy whole foods affordable in our current system. The crucial feature of the co-op is that it is

---

* My alma mater, the University of California, Santa Cruz, manages the Alan Chadwick Garden, which he founded in 1967 on the principles of French intensive/biodynamic farming, thereby popularizing these farming methods in the United States after industrial farming had taken over.

entirely member owned and operated. This means that every member must work a mandatory 2.75-hour shift once a month. As member-worker-owners, we are entitled to shop at the co-op at prices about 20 percent above wholesale prices—in comparison to most grocery stores that charge a 100 percent markup. This significant savings allows many members to afford healthy, organic, and regionally produced foods that otherwise would be out of reach—and these foods include not only an abundance of high-quality organic produce, nuts, seeds, and grains in bulk but also locally sourced animal products like grass-fed beef and milk, yogurt, and cheeses from grass-fed cows; eggs and chickens raised on pasture; and locally caught wild fish.

Established in 1973, the co-op now has more than seventeen thousand members, sells roughly one million dollars' worth of food a week, and consistently runs an annual profit—so much so that we have been able to lend money to new food co-ops around the world that want to replicate our model. The co-op is run as a collective and has monthly general meetings at which members vote on all decisions. This kind of direct participatory decision making is a superb model for better ways to run our lives and supply our crucial day-to-day needs. The co-op also enjoys significant buying power in its dealings with farms and food suppliers around the country, which helps to keep costs down and removes the middleman, who profits from both farmers and consumers.

Hepworth Farm in Upstate New York is growing food specifically for the co-op. Allen Zimmerman, who was the food co-op's main produce buyer until his retirement in 2015, told me that the co-op and farm grew together and that the relationship allowed both to succeed more than they could have separately. "They would have grown, we would have grown, but boy did we do it well because we did it together," Zimmerman said. Hepworth grew pesticide-free foods for the co-op before there was organic, Zimmerman said, and when the time came, he talked Hepworth into getting certified as organic, which can be time consuming and expensive. Since then, Hepworth has grown to a few hundred acres and buys more farmland every year. "[Hepworth] made our produce part of what people talk about. Most farms don't grow more than a dozen commodities, and [Hepworth] grow[s] over a hundred. And they also do something that most farms can't do—they grow for us. We say, 'Can we have Shiso?' And they say, 'Mid-September.' They really are so responsive."[31]

The importance of these relationships and direct communication between the members of the co-op and local farmers cannot be overstated, and it is sorely missing from mainstream grocery stores across the country. Indeed farmers, food producers, and consumers all suffer when the corporate entities, whether the grocery stores or the suppliers, do the negotiating to enhance their bottom line rather than working in the interest of the consumers or the farmers. Still, there is something else about the co-op that could benefit more communities—the feeling of involvement and cooperation removes the sense that you are just a consumer in someone else's space. It is hard to quantify this shift in experience, but it creates an awareness about our current consumer-based society that you won't find in the aisles of Whole Foods or Safeway. Joe Holtz, one of the founding members of the co-op, put it this way: "The culture in the co-op of members/owners stepping forward to do their share of the work with an attitude and approach that on average includes caring and a feeling of some connection and belonging is the culture that has sustained the co-op and is the major secret to our success," he said in 2016.[32]

The growth of the co-op from just a few dozen members at its founding to more than seventeen thousand members today while maintaining this culture is quite remarkable and speaks to what people can do when they come together because of their shared interest in access to and the affordability of healthy foods. And although Park Slope is an affluent area, the co-op's members are diverse by every measure, and many travel significant distances throughout New York City and beyond to be members. It's also worth noting that when the co-op began back in 1973, Park Slope was not a particularly desirable neighborhood. "The roots of the co-op go back to well before gentrification took a strong hold in this area of Brooklyn. . . . Grocery shopping is local and the locals have gotten more well-to-do. But the co-op pulls from far afield. . . . We have a wide diversity of income levels," Holtz said.

The collective buying power that the co-op enjoys is also crucial; when the co-op decides it does not want to continue to support a particular food or a particular brand, suppliers and companies take notice. This is quite different from the food movement's usual refrain: to pressure food companies and corporations to change their practices through our individual buying power. The model of the Park Slope Food Coop provides a way to leverage buying power and influence food companies more profoundly. The co-op has voted to boycott various companies over the years, including Coca-Cola, Bumble Bee tuna, Poland Spring water, and Domino Sugar because the members agreed that these companies' products and practices did not align with the co-op's mission statement, which in part reads: "We offer a diversity of products with an emphasis on organic, minimally processed and healthful foods. We seek to avoid products that depend on the exploitation of others. We support non-toxic, sustainable agriculture."

Large-scale boycotts are far different from the food movement's advocacy of voting with our dollars. In many cases, pressure from consumers results in only superficial victories that only perpetuate the destructiveness of the larger industrial system. These changes are often meaningless and sometimes actually strengthen the opposition. If we are satisfied with small tweaks to existing industrial products, the system continues to run and profit as is. For example, in April 2016, Walmart announced that it would only sell eggs from cage-free chickens beginning in 2025. Joel Salatin, a farmer, writer, and activist, relayed an important point in a Facebook post following Walmart's announcement. He said that instead of changing its suppliers to local and regional farmers who are already raising chickens humanely and sustainably and could immediately supply Walmart with eggs, the company was going to retain its suppliers but give them time to adjust the way they house the chickens—hence the 2025 deadline. Yet we know that "cage free" means next to nothing if huge suppliers own the chickens. It means only that chickens are crammed into open-air barns, and each chicken has one square foot of space as specified by the regulations on cage free. The chickens still do not have access to the outdoors, which means no pasture, no fresh air, and no sunshine. Yet Walmart's self-congratulatory announcement will lull consumers into the belief that significant change is afoot. Or, as Joel Salatin wrote, "The biggest problem with this is that thousands of people will feel like they've arrived at animal welfare nirvana, which will make consumers more lethargic about seeking out the real deal."

Such announcements can trick consumers into thinking they have done the right thing based on marketing and label claims and not much else. Sadly the food movement has been integral to the creation of the entire industrial organic food industry, which profits off the cursory knowledge of many consumers about food, health, and animal welfare. We must remember that tweaking around the edges of an unsustainable food industry is a grave tactical mistake. *Negotiation cannot replace mobilization.*[33]

Can we agitate to radicalize the existing food movement? That will be a key question as we move forward because, despite my criticisms of this movement, many people operating under the rubric of the food movement have helped to effect change. Indeed, this book builds upon the awareness around food and health that the movement has already raised. Even those I criticize for their shortcomings have had a large influence on how Americans now view and talk about food. But the food movement is still too narrowly defined and not inclusive enough.

That is why it is crucial that we continue to change the culture around food and that we continue to bring more and more people into the fold. This will mean that you, the readers, must demand better from our government and our food producers. It also means that you need to reach out to people across all socioeconomic, political, and geographic barriers to stress that a healthy food system directly benefits all of us. One way to frame this is that it is about protecting the young. Feeding our children healthy foods, breast-feeding them, and teaching them cooking skills has become nothing short of political work. This is work that protects our children and serves an enormous public health interest. Selma James made the important point that, given how our culture functions today, there is no sense that protecting the young is a crucial part of what our society should be doing. "Women are fed up and exhausted in industrial countries, and they want another kind of life and another kind of recognition, not only the recognition of pay equity, which is tremendously important, but also the recognition for caring as the priority," she said.[34] Wages for Housework and my proposed program, Pay People to Cook at Home, can help reframe the importance and value in the work of raising children and all that it entails—especially the work of preparing healthy meals for our families. James told me that men's ability to spend time with the family and to acquire and use skills like cooking needs to become a priority as well. "Wages for Housework represents an anticapitalist framework in the sense that, instead of the market being central, which we all are slaves to, people have to be central and the market must answer our human needs rather than the other way around," she said. "People have never even considered that they could organize their own lives in a collective way."[35]

Both mothers and fathers will need to be engaged with their families and engaged with the important skills of cooking for their families. Remember that this is not new—men, women, and children all were involved in and responsible for household work until fairly recently in our history. Yet we also know that the challenge moving forward will be to create a society in which actual family values matter, and household work, especially cooking, will not be based on regressive gender roles.

Collective action is also vitally important for demanding a say in the production of our food. Without food autonomy, creating change will be difficult. This does not mean we all need to grow our own food, but we do need to engage with our food supply and support those who do the growing. A large part of that engagement will be direct action protests to demand that Big Food and Big Ag respond to the needs of people before profit and stop producing harmful foods that are reliant on harmful chemicals. The reemergence of protest across the country (and world) is encouraging and a sign of an engaged populace fed up with the status quo. The protests at Standing Rock, the Black Lives Matter movement, and the counterprotests that have emerged in response to white nationalists and neo-Nazis serve as key examples of the tactical importance of direct action in the current political climate. It is becoming ever more apparent that we will need to organize and resist the status quo for real change in our food supply.

Member-owned and -run food cooperatives are an excellent model for collectivizing, and the Park Slope Food Coop provides a blueprint for doing so successfully. What other creative cooperative ideas can we come up with? How can we collectivize housework, cooking, and child care? These are questions we must seriously engage if we want to change our day-to-day lived realities.

Here is what we need to remember moving forward:

- Calls for better access to food and food education do nothing to address the predatory nature of Big Food and Big Ag marketing and the normalization of their unhealthy products.
- Calls to buy organic do nothing to alter the production and use of dangerous chemicals that lace the majority of our foods.

- Providing recipes for quick and easy-to-prepare meals does not address the problems of gender oppression, poverty, overwork, and too little pay.
- Pushing the food industry to make minor changes often undermines the possibilities for real, significant change.

The vast changes to our food system, and by extension our bodies, represent one of the biggest problems that we as a species must confront. Environmental and public health disasters are already upon us, and continuing on our current path is no longer an option. What we need is nothing short of a societal transformation—placing human needs above the needs of the market—and true human needs for good health are part of preserving and protecting the natural world. Adrian Parr, a professor of environmental politics at the University of Cincinnati, has said that climate degradation is akin to crimes against humanity, and one of the biggest perpetrators is the fossil fuel industry. "A crime against humanity is an action that causes severe and unnecessary human suffering, and environmental destruction unquestionably degrades the quality of human life," she said.[36] I would expand that definition to include the destruction of our food supply, which is ultimately destroying our bodies. Thought of in that way, Big Food and Big Ag companies are guilty of the same kinds of crimes against humanity as the fossil fuel industry—which, not coincidentally, enables both Big Food and Big Ag to function as they do.

Our efforts to "fix" the food system as it is will be futile—buying organic, pressuring food companies through what we buy individually, or making healthier individual choices to avoid various toxic environmental chemicals will never bring about real change. If using the existing framework of the current system to effect change is not possible, the only other option is to work together and demand it through direct action protests, large-scale boycotts, and rebuilding our communities. And we must do so while also figuring out how to restructure our day-to-day lives to include growing, producing, distributing, cooking, and consuming our foods, as well as feeding and interacting with our families and communities. We must begin to act as if our lives depend on it.

Conclusion

Protecting Our Bodies and Our Future

Our bodies are porous to all that is around us, absorbing and assimilating everything in our environment at once. With the onslaught of environmental chemicals that in many cases are toxic to all forms of life, we are asking our bodies to deal with a lot. At the same time, we are not providing our bodies with the necessary nutrients they need to be optimally healthy. This is a dangerous combination as demonstrated by the unprecedented increase in our rates of diet-related disease. We are also subject to cultural forces that have normalized industrial food and separated us from food production and, by extension, each other.

Part of changing the way we approach our food system will entail changing the way we view our world, and this means working together in collective action—change will not come from individual choices made within the context of our current economy. We will need massive policy reforms in combination with organizing from the ground up. A reformulation of the way we grow, produce, and consume foods can be seen as a continuation of the civil rights movement, the feminist movement, and the overall struggle for economic justice. Ultimately the battles about how and what we eat are about values. What do we value in our current food system? Speed, efficiency, convenience, profit? What about public health, nourishment, stewardship of the land and water, the preservation of Earth and all its species, and the protection of the young and their future?

Some critics say that the dangers of the food system in relation to our health are overblown and point to the continued increase in average life span in this country as proof. But as I've mentioned before, as of December 2016 that argument is no longer valid. The National Center for Health Statistics announced that for the first time since 1993 life expectancy had declined in the United States, and six of the ten leading causes of death are directly related to our diets: heart disease, cancer, stroke, Alzheimer's disease, diabetes, and kidney disease.[1] And perhaps the most startling shift of all has occurred among the young in our population—rates of childhood obesity, diabetes, and even heart disease are surging, and for the first time in two centuries children are expected to have shorter life spans than their parents.[2]

It is obvious that we need to massively overhaul our food and our food system to protect public health. Using food as the central rallying point for organizing has the potential to bring powerful change—maybe in ways that no other movement does. A political movement to create a healthy, just, and sustainable food supply will be the foundation for making our societal systems work for human beings with the environment, not for the free market at the expense of the environment. Organizing around food is also incredibly unifying because the need for good healthy food is universal. And there is one other thing about food: eating is pleasurable. Healthy and delicious food has always been and will always be a way to bring people together; it is encoded in our biology. That means a radicalized and politicized food movement has limitless potential for success because it questions the very foundations of our current exploitative societal structures from a unifying perspective.[3]

So here's a call to action, a food manifesto that can guide us as we move toward creating a healthy, just food system that will benefit human health and the environment. We need new and innovative ideas for organizing and building alternative models in our communities, *and* we need to stand together in opposition to Big Food and Big Ag. We must demand high-quality and safe foods for all of us. In this way, we will protect not only our own health but the environment, the young, and future generations. Handing our food supply over to the free market in the last century has resulted in unprecedented destruction to our environment and to our health. We have to take our food back.

A NEW FOOD MOVEMENT MANIFESTO

Stop the predatory marketing of poor-quality industrial foods, particularly to children and people of color, who are unfairly targeted, and stop celebrities from marketing these foods.

The food industry follows the marketing playbook of the tobacco industry quite closely. When the tobacco industry saw its sales to adults were slowing because they understood the health effects of smoking, it targeted children and teens with much of its advertising, often airing commercials during popular children's TV shows. (In 1998 Big Tobacco reached a settlement with forty-six state attorneys general in which it agreed to greatly curtail its advertising to children and teens.[1]) The food industry similarly targets youth, even toddlers as young as two, with its marketing. Like the food industry, the tobacco industry also targeted communities of color and placed more ads in those neighborhoods than elsewhere. The tobacco industry also used movie stars and athletes to hawk cigarettes, whereas today we have Beyoncé selling Pepsi, Taylor Swift selling Diet Coke, and LeBron James doing commercials for McDonald's. Recognizing the threat to public health, in 1970 President Richard Nixon signed a law that banned the advertising of cigarettes on television and radio.[2] Then in 1997, the Tobacco Master Settlement Agreement banned billboard and public transportation advertising, as well as ads that targeted young people in most states.[3] The food and beverage industry should be banned from these kinds of practices as well.

Stop the marketing of infant formula to parents.

Advertising of infant formula, the first processed food, should also have tight restrictions. With what we now know about the importance of breast milk for babies' immediate and long-term health, parents should turn to infant formula only when it is medically necessary. Although most hospitals prohibit the practice of handing new parents "goody bags" filled with formula samples, this practice should be formally banned, and we should also limit the kinds of advertising formula manufacturers use to reach new parents. Furthermore, the misleading labels on formula packaging should also be banned.

Place warning labels on all industrial food packaging: "These foods may be harmful to your health."

Warning labels would be much more effective than the nutrition panels and calorie counts that packaged foods carry today. Most consumers don't know what to make of all that information, and packaging labels are misleading. A simple warning label, much like what is now required for cigarettes, should be prominent on industrial food packaging. The European Union already does this for particular ingredients, like certain food dyes. The label reads: "May have an adverse effect on activity and attention in children." We could start by assembling an independent body of experts to determine the kinds of ingredients in industrial foods that warrant a warning label and what the label should say. Some of the worst offenders of the industrial food world should bear the simple label, "These foods may be harmful to your health."

Stop the use of the thousands of chemicals in and on our food supply that independent, third-party testing has not proved safe.

No agency or law now ensures the safety of the chemicals in and on our food supply, which presents a grave threat to public health. We should create a completely independent, third-party verification system, based on the latest, most relevant science, that would rigorously test all chemicals that manufacturers wish to add to our food supply.

Demand easy and affordable access for all to chemical-free, healthy whole foods by creating a federal urban farm program.

By creating a federal urban farm program, we could provide whole foods in many areas of the country where they are hard to find. Improvements to public health could be substantial if the United States inaugurated a program based on urban farms, which could produce large quantities of affordable, healthy food, employ local community members, and provide food and environmental education hubs. In 2016 Senator Debbie Stabenow, a Michigan Democrat, introduced the Urban Agriculture Act to extend the U.S. Department of Agriculture's farm loan programs to urban farmers, support farm co-ops, invest in urban agricultural research, and help improve access to fresh local foods. It has never made it out of committee.[4] The bill is a step in the right direction but should be taken further if we hope to remedy our national health crises.

Demand nutrition and cooking education in all public schools.

We need a comprehensive curriculum in all public schools to teach the fundamentals of whole food nutrition (not those based on industry guidelines) and the basics of cooking and preparing whole foods from scratch. Many schools could also have a kitchen garden and use it to teach children about growing vegetables and herbs. Such hands-on learning experiences are invaluable and could easily be linked to science courses, making them relevant to students' everyday lives.

Demand a universal basic income.

Nothing will change if most people do not have enough money to buy healthy foods. Having a basic income would mean that people would not have to choose between paying the rent, paying the bills, paying for medications, or eating healthy foods. Food is often the lowest priority on that list because it offers some flexibility, but that means people then buy the least expensive options—

fast food and convenience foods. This is a vicious cycle because poor food intake leads to myriad health issues, which eventually exact their own financial burden and suffering. According to recent research, having a basic income does not reduce the labor supply or encourage people to drop out of the labor force. Recipients also spend more money on higher-quality foods. In Canada the long-term tracking of providing citizens with a basic income shows that it results in healthier citizens, fewer hospitalizations, improved mental health outcomes, and improved school attendance, grades, and test scores.[5] A basic income will also allow workers to reject the lowest-paying jobs in the agricultural and food industries, like those in slaughterhouses or farm work that entails dangerous pesticide exposure. Until everyone has equal access to healthy whole foods, we must provide money for people to buy them. Affording healthy food should be a right and not a privilege.

Demand payment for cooking and other household work.

To some this may seem like the most unrealistic idea, yet it is the most important. Gender inequality is one of the most intractable problems in our society, with women still earning only a fraction of what men make in the workforce, and even that fraction is significantly smaller for women of color. Ninety percent of all caretakers are women, and most are also employed outside the home. To truly address our national public health crises, we have to begin in the home. Without at least one parent with the time to shop, prepare, cook, and clean up after meals, telling people to cook more and eat healthy is futile. People must be financially able to do so. Therefore, a program that would pay people to cook at home, especially while taking care of small children, is crucial to improving our individual health and the health of our society.

Demand a federal policy that requires employers to provide a minimum of six months' paid parental leave, ensuring the option to breast-feed as a right.

Countries around the world offer both maternal and paternal leave, with great benefits to public health. In fact, the United States is the only industrialized nation that does not provide guaranteed paid parental leave. All major health agencies around the world, including the American Academy of Pediatrics and the World Health Organization, agree that a minimum of twelve months of breast-feeding is optimal for the health of both mother and baby. But in our current economy, many women who want to breast-feed are unable to because they would lose money they depend on. Because babies require only breast milk for all their nutrition needs during their first six months, six months' maternity leave should be a basic right.

ACKNOWLEDGMENTS

The research for this book began more than ten years ago when I started researching, writing, and thinking seriously about food, health, nutrition, and our food culture. There are many people who played a role in the development of this book to whom I owe a debt of gratitude. Early on I was fortunate to learn a lot from the very good nutrition teachers Laura Knoff, Nori Hudson, and Ed Bauman.

Thank you to Amy Seek for introducing me to Paula Crossfield, then the editor of *Civil Eats*. Working with Paula as well as Naomi Starkman helped launch my writing career and I am grateful for their support. In the book, I build on ideas from my *Civil Eats* articles and I appreciate the platform the site has provided me over the years. I would also like to thank Trish Hall, then an editor at *The New York Times*, for her help with two important op-eds that were part of the foundation for much of the thinking in this book. Thanks also to Corby Kummer at *The Atlantic*, and to the editors at *Ecocentric*, *Grist*, and the *Huffington Post*. Kate Lowenstein at *VICE*, who is a pleasure to work with, generously read an excerpt of the book even when she had virtually no time to spare. Ideas I first published in one or another of these outlets are central to my argument and appear in various places in the book.

I'd also like to thank my editors at St. Martin's Press, Karen Wolny and Elisabeth Dyssegaard. Both came to the project with enthusiasm and encouragement. Karen's guidance about the writing and her excitement about the project no doubt improved the book. Elisabeth's expert shepherding of the book through the publication process was impeccable. Thanks also to everyone who contributed to the production of this book at St. Martin's Press.

Sonjia Hyon eagerly and smartly fact-checked several chapters and double-checked citations and research, even on very short notice. Thanks to the talented photographer, my friend Carin Backoff. And thanks to Erika Camplin, Erica Wides, and Edith Lee for their unfailing support and advice.

I want to thank my former professor, mentor, and friend, Louis Chude-Sokei, who believed in my work since my days as an undergraduate at UC Santa Cruz, was instrumental to my research as a graduate student, and encouraged me throughout the writing of this book. He also generously read chapters even in the midst of his own work and writing.

My agent, Jill Kneerim, has been amazing, going above and beyond beginning with the development of the proposal all the way through the marketing of the book. She truly believes in the project and in my work. Her steady presence guiding me through the many ups and downs of book writing has been invaluable. Thanks also to Lucy Cleland and the other fine people at Kneerim & Williams.

Thanks to Mark Bittman, who has always been supportive of my work and who generously provided mentorship and encouragement along the way. Thanks to Steve Fraser for helping with the early development of this project and providing the crucial introduction to my agent, Jill. And thank you to Chuck D, Laurie David, Dennis Goodman, Mary Esther Malloy, and David McCarron for their support.

I am indebted to the important work of Selma James, founder of the Wages for Housework Campaign, which informs my own thinking and writing. I've been lucky to work with Selma and her colleague Phoebe Jones over the years.

I've conducted hundreds of interviews with scientists and researchers and I would like to thank each one who appears in this book for generously giving their time. I would especially like to recognize the contributions of Bruce Blumberg, Bruce German, Andrea Gore, Robert Lustig, Maricel Maffini, Fred vom Saal, Laura Vandenberg, and Tom Zoeller—not only for the hours upon hours we spent talking—but for the important work they do in the service of the public interest.

My family has been a source of great inspiration. My Bubie and Zayda, Dorothy and Herman Wartman, and my great uncle, Ben Friedman, put me on the path to understanding the importance and value of cooking, eating well, and its connection to health from the moment I sat down at their dinner table. My Grandaddy, John Hobe, instilled in me a love for nature, gardening, and fresh food. Both he and my grandmother, Barbara Hobe, showed me the value in cooking with local ingredients. My mom, Barbara Hobe Wartman, and my dad, Paul Wartman, continued these traditions in valuing health and good food and I was lucky to have a mom who cooked a healthful dinner nearly every night.

My mom, dad, and sister, Dr. Sarah Wartman, were also instrumental to the research and writing of the book. They each read and reread versions of chapters, helped me flesh out ideas and concepts, and always supported me and my work with enthusiasm. I especially want to thank my mom, who has always been my first editor, and who read countless drafts, providing valuable insight and careful editing.

Last but not least I want to thank my husband, Brian, who quite literally made the writing of this book possible, giving me the freedom to research and write over the course of a decade. We had a baby as this book was nearing completion, and so I finalize this project with great hope and determination that we will fight for a safer, healthier world for him, his generation, and all those still to come.

NOTES

CHAPTER ONE: OUR INDUSTRIAL FOOD LANDSCAPE

1. Kristin Wartman, "Big Food Co-Opts Nutrition Group's Message," *Civil Eats*, January 24, 2013, http://civileats.com/2013/01/24/new-report-corps-co-opt-nutrition-groups-message/.
2. "Sponsor Thank You," *The American Heart Association*, http://www.heart.org/HEARTORG/General/Sponsor-Thank-You_UCM_469280_Article.jsp# (accessed 12/12/17).
3. American Heart Association Heart-Check Food Certification Program Application Packet, http://www.heart.org/idc/groups/heart-public/@wcm/@fc/documents/image/ucm_447605.pdf (accessed 12/28/17).
4. Luz Varela, email to author, August 22, 2017.
5. Nina Planck, "Just the Fats," *Bon Appetit*, April 6, 2008, https://www.bonapetit.com/trends/article/just-the-facts; Mahsid Dehghan et al., "Association of Fats and Carbohydrate Intake with Cardiovascular Disease and Mortality in 18 Countries from Five Continents (PURE): A Prospective Cohort Study," *The Lancet* 390, no. 10107 (August 2017): 2050–62.
6. Nina Planck, *Real Food* (New York: Bloomsbury, 2006), 212–13.
7. Michael Murray, *The Encyclopedia of Healing Foods* (New York: Atria Books, 2005), 612–13; Planck, *Real Food*.

8. U.S. Department of Health and Human Services and U.S. Department of Agriculture, *Dietary Guidelines for Americans, 2015–2020*, 8th ed., December 2015, https://health.gov/dietaryguidelines/2015/resources/2015-2020_Dietary_Guidelines.pdf.
9. Benoit Chassaing et al., "Corrigendum: Dietary Emulsifiers Impact the Mouse Gut Microbiota Promoting Colitis and Metabolic Syndrome," *Nature* 519 (2015): 92–96.
10. S. Jay Olshansky et al. "A Potential Decline in Life Expectancy in the United States in the 21st Century," *The New England Journal of Medicine* 352 (March 17, 2005): 1138–45.
11. Jiaquan Xu et al. "Mortality in the United States, 2015," Centers for Disease Control and Prevention, National Center for Health Statistics Data Brief No. 267 (December 2016).
12. Natasha Gilbert, "One-Third of Our Greenhouse Gas Emissions Come from Agriculture," *Nature* 10 (October 31, 2012): 1038.

CHAPTER TWO: WHAT ARE WE ACTUALLY EATING?

1. Anahad O'Connor, "Study Questions Fat and Heart Disease Link," *The New York Times*, March 17, 2014; Pavel Grasgruber et al., "Food Consumption and the Actual Statistics of Cardiovascular Diseases: An Epidemiological Comparison of 42 European Countries," *Food and Nutrition Research* 60 (2016): 31694; Mahsid Dehghan et al., "Association of Fats and Carbohydrate Intake with Cardiovascular Disease and Mortality in 18 Countries from Five Continents (PURE): A Prospective Cohort Study," *The Lancet* 390, no. 10107 (August 2017): 2050–62.
2. Douglas Main, "BPA Is Fine, If You Ignore Most Studies About It," *Newsweek*, March 15, 2015.
3. Z. Chun Yang et al., "Most Plastic Products Release Estrogenic Chemicals: A Potential Health Problem That Can Be Solved," *Environmental Health Perspectives* 119, no. 7 (July 2011): 989–96.
4. Kristin Wartman, "Not Your Grandma's Milk," *Grist*, September 13, 2011, http://grist.org/scary-food/2011-09-12-not-your-grandmas-milk/full/.
5. Fred Kummerow, "Interaction Between Sphingomyelin and Oxysterols Contributes to Atherosclerosis and Sudden Death," *American Journal of*

Cardiovascular Disease 3, no. 1 (February 2013): 17–26; Fred Kummerow, interview by author, April 10, 2015, Urbana, Illinois.

6. Wartman, "Not Your Grandma's Milk."
7. For a more exhaustive history of Keys and the cholesterol hypothesis, see Nina Teicholz's book *The Big Fat Surprise* (New York: Simon & Schuster, 2014).
8. David A. McCarron, interview by author, July 24, 2017, Davis, California.
9. "The Fat of the Land," *Time*, January 13, 1961, vol. 77, no. 3, pp. 48–56, http://content.time.com/time/magazine/article/0,9171,828721,00.html.
10. Teicholz, *Big Fat Surprise*, 231; McCarron, interview.
11. Kummerow, interview.
12. Mary Enig, who died in 2014, was another important researcher who raised concerns about trans fats (and extolled the virtues of saturated fats).
13. Walter Willet, "The Scientific Case for Banning Trans Fats," *Scientific American*, December 13, 2013.
14. Nina Planck, *Real Food* (New York: Bloomsbury, 2006), 200.
15. Teicholz, *Big Fat Surprise*, 275; M. Ash, *Vegetables Oils: Supply, Disposition and Utilization Tables, 1912–1965*, Report of Inter-Society Commission for Heart Disease Resources (Washington, DC: U.S. Department of Agriculture, 2012); Kummerow, interview.
16. Sheila Innis and Russell W. Friesen, "Essential n–3 Fatty Acids in Pregnant Women and Early Visual Acuity Maturation in Term Infants," *American Journal of Clinical Nutrition* 87, no. 3 (March 7, 2008): 548–57; M. R. Sanz Sampelayo et al., "Thermogenesis Associated to the Intake of a Diet Non-Supplemented or Supplemented with n-3 Polyunsaturated Fatty Acid-Rich Fat, Determined in Rats Receiving the Same Quantity of Metabolizable Energy," *Annals of Nutrition and Metabolism* 50, no. 3 (2006): 184–92.
17. Charles Benbrook, email to author, December 7, 2013; William Smith, "Anti-inflammatory Effects of Omega 3 Fatty Acid in Fish Oil Linked to Lowering of Prostaglandin," paper presented at the meeting of the Federation of American Societies for Experimental Biology, April 4, 2006, San Francisco.

18. Mikael Eriksson et al., "Pesticide Exposure as Risk Factor for Non-Hodgkins Lymphoma Including Histopathological Subgroup Analysis," *International Journal of Cancer* (October 1, 2008): 1657–63; Siriporn Thongprakaisang et al., "Glyphosate Induces Human Breast Cancer Cells' Growth via Estrogen Receptors," *Food and Chemical Toxicology* 59 (September 2013): 129–36; R. M. Romano et al., "Prepubertal Exposure to Commercial Formulation of the Herbicide Glyphosate Alters Testosterone Levels and Testicular Morphology," *Archives of Toxicology* 84, no. 4 (April 2010): 309–17.
19. John Peterson Meyers et al., "Concerns over Use of Glyphosate-Based Herbicides and Risks Associated with Exposures: A Consensus Statement," *Environmental Health* 15 (February 17, 2016).
20. Michael Wines, "Monarch Migration Plunges to Lowest Level in Decades," *The New York Times*, March 14, 2013, http://www.nytimes.com/2013/03/14/science/earth/monarch-migration-plunges-to-lowest-level-in-decades.html.
21. Kristin Wartman Lawless, "No One Knows How Much Herbicide Is in Your Breakfast," *Vice*, May 2016, https://www.vice.com/en_us/article/no-one-knows-how-much-herbicide-is-in-your-breakfast.
22. Pollak Holmes et al., "Dietary Correlates of Plasma Insulin-like Growth Factor I and Insulin-like Growth Factor Binding Protein 3 Concentrations," *Cancer Epidemiology, Biomarkers, and Prevention* (September 2002): 852–61; Stampfer Chan et al., "Plasma Insulin-like Growth Factor-I and Prostate Cancer Risk: A Prospective Study," *Science* (January 1998): 563–66; Jin Yu et al., "Insulin-like Growth Factors and Breast Cancer Risk in Chinese Women," *Cancer Epidemiology, Biomarkers, and Prevention* (August 2002): 705–12.
23. Bruce Blumberg, interview by author, July 16, 2014, Irvine, California.
24. Pagan Kennedy, "The Fat Drug," *The New York Times*, March 8, 2014, https://www.nytimes.com/2014/03/09/opinion/sunday/the-fat-drug.html?_r=0; Martin Blaser, "Why Antibiotics Are Making Us All Ill," *The Guardian*, June 1, 2014, https://www.theguardian.com/society/2014/jun/01/why-antibiotics-making-us-ill-bacteria-martin-blaser.
25. Blumberg, interview; Dr. Jane Muncke, interview by author, July 17, 2014, Zurich, Switzerland.

26. Andreas Kortenkamp et al., "Synergistic Disruption of External Male Sex Organ Development by a Mixture of Four Antiandrogens," *Environmental Health Perspectives* 117, no. 12 (December 2009): 1839–46; Kristin Wartman, "Our Deadly Daily Chemical Cocktail," *Civil Eats*, April 26, 2011, https://civileats.com/2011/04/26/our-deadly-daily-chemical-cocktail/.

CHAPTER THREE: LOSING OUR FOOD ILLUSIONS

1. Michael Pollan, "Naturally," *The New York Times Magazine*, May 13, 2001, http://www.nytimes.com/2001/05/13/magazine/naturally.html
2. Fred Kummerow, interview by author, April 10, 2015, Urbana, Illinois.
3. Reuters Staff, "Danone Raises 2017 EPS Guidance after WhiteWave Acquisition," *Reuters*, April 20, 2017, https://www.reuters.com/article/danone-results/danone-raises-2017-eps-guidance-after-whitewave-acquisition-idUSFWN1HS0Q4.
4. Cornucopia Institute, https://www.cornucopia.org/2014/02/horizon-organic-factory-farm-accused-improprieties/.
5. Sara Loveday, spokesperson for WhiteWave Foods, email to author, July 13, 2015.
6. Kristin Wartman, "Not Your Grandma's Milk," *Grist*, September 13, 2011, http://grist.org/scary-food/2011-09-12-not-your-grandmas-milk/full/.
7. Charles M. Benbrook et al. "Organic Production Enhances Milk Nutritional Quality by Shifting Fatty Acid Composition: A United States-Wide, 18-Month Study," *PLOS ONE* 8(12): e82429 (December 9, 2013).
8. Cynthia A. Daley et al., "A Review of Fatty Acid Profiles and Antioxidant Content in Grass-Fed and Grain-Fed Beef," *Nutrition Journal* 9, no. 10 (March 10, 2010), https://doi.org/10.1186/1475-2891-9-10.
9. Kristin Wartman, "Organic Agriculture: Fifty (Plus) Shades of Gray," *Ecocentric* (blog), Grace Communications Foundation, September 18, 2012, http://www.gracelinks.org/blog/1207/organic-agriculture-fifty-plus-shades-of-gray.
10. Donald R. Davis, et al., "Changes in USDA Food Composition Data for 43 Garden Crops, 1950 to 1999," *Journal of the American College of Nutrition* 6 (December 2004): 669–82.
11. Crystal Smith-Spangler, MD, et al. "Are Organic Foods Safer or Healthier

Than Conventional Alternatives? A Systematic Review," *Annals of Internal Medicine*, 157, no. 5 (September 4, 2012): 348–66.

12. Asa Bradman et al., "Effect of Organic Diet Intervention on Pesticide Exposures in Young Children Living in Low-Income Urban Agricultural Communities," *Environmental Health Perspectives* 123, no. 10 (October 2015): 1086–93.
13. Kirsten Brandt et al., "Agroecosystem Management and Nutritional Quality of Plant Foods: The Case of Organic Fruits and Vegetables," *Critical Reviews in Plant Sciences* 30, nos. 1–2 (2011): 177–97; Lynne Peeples, "Stanford Organic Study: Have Faulty Methods, Political Motivations Threatened Kids' Health?," *Huffington Post*, September 13, 2012, https://www.huffingtonpost.com/2012/09/13/stanford-organics-study-public-health_n_1880441.html.
14. Fernando Vallejo et al., "Health-Promoting Compounds in Broccoli as Influenced by Refrigerated Transport and Retail Sale Period," *Journal of Agricultural and Food Chemistry* 51, no. 10 (2003): 3029–34.
15. Penn State Extension, Broccoli Production, August 14, 2017, https://extension.psu.edu/broccoli-production (accessed January 7, 2018); Brian Palmer, "The C-Free Diet," *Slate*, July 10, 2013, http://www.slate.com/articles/health_and_science/explainer/2013/07/california_grows_all_of_our_fruits_and_vegetables_what_would_we_eat_without.html
16. Ray Bradley, interview by author, September 3, 2012, Brooklyn, New York; Stephen Goff et al., "Plant Volatile Compounds: Sensory Cues for Health and Nutritional Value?," *Science* 10, no. 311 (February 2006): 815–19.
17. Katherine Koh, "One-Third of Homeless Are Obese," *Journal of Urban Health* 89, no. 6 (May 2012): 952–64; "Share of Milk Sales in the US in 2016, by Category," *Statista*, http://www.statista.com/statistics/249077/share-of-us-milk-sales-by-category/ (accessed March 17, 2016).
18. Sarah Holmberg and Anders Thelin, "High Dairy Fat Intake Related to Less Central Obesity: A Male Cohort Study with 12 Years' Follow-Up," *Scandinavian Journal of Primary Health Care* 31, no. 2 (June 2013): 89–94; Mario Kratz et al., "The Relationship Between High-Fat Dairy Consumption and Obesity, Cardiovascular, and Metabolic Disease," *European Journal of Nutrition* 52, no. 1 (February 2013): 1–24; Alison Aubrey, "The

Full-Fat Paradox: Whole Milk May Keep Us Lean," *NPR*, February 12, 2014, http://www.npr.org/sections/thesalt/2014/02/12/275376259/the-full-fat-paradox-whole-milk-may-keep-us-lean.

19. Paul Lips, "Worldwide Status of Vitamin D Nutrition," *The Journal of Steroid Biochemistry and Molecular Biology* 121, nos. 1–2 (February 21, 2010): 297–300; Adit Ginde et al., "Demographic Differences and Trends of Vitamin D Insufficiency in the US Population, 1988–2004," *The JAMA Network: Archives of Internal Medicine* 169, no. 6 (March 23, 2009): 626–32.
20. New York State Department of Health, "Folic Acid: The Vitamin That Helps Prevent Birth Defects," April 2007, https://www.health.ny.gov/publications/1335/ (accessed April 1, 2016).
21. George Mateljan Foundation, "Folate," The World's Healthiest Foods, no date, http://www.whfoods.com/genpage.php?tname=nutrient&dbid=63 (accessed April 3, 2016); Royal DSM, *Integrated Annual Report for 2016*, http://annualreport.dsm.com/ar2016/en_US/index.html.
22. M. F. K. Fisher, *How to Cook a Wolf* (New York: Duell Sloan Pearce, 1942), 73.
23. Shane Starling, "World's Fastest Growing Functional Food in 2013? Infant Formula (By a Mile)," *NUTRAingredients*, October 30, 2013, https://www.nutraingredients.com/Article/2013/10/30/Infant-formula-is-world-s-fastest-growing-functional-food-in-2013.
24. "Retail Value of the Baby Milk Formula Market Worldwide in 2010 and 2020," Statista (accessed January 4, 2017), https://www.statista.com/statistics/719436/global-market-size-baby-formula/.
25. David R. Jacobs et al., "Food Synergy: An Operational Concept for Understanding Nutrition," *The American Journal of Clinical Nutrition* 89, no. 5 (May 2009): 1543–48.
26. James Meikle and Luke Harding, "Denmark Bans Kellogg's Vitamins," *The Guardian*, August 12, 2004, https://www.theguardian.com/world/2004/aug/12/foodanddrink.
27. George F. M. Ball, *Vitamins: Their Role in the Human Body* (Oxford: Blackwell Publishing, 2004), 223.
28. Environmental Working Group, *How Much Is Too Much? Excess Vitamins and Minerals in Food Can Harm Kids' Health*, June 19, 2014, https://www.ewg.org/research/how-much-is-too-much#.Wrqcg4jwaUk.

29. Ibid.
30. "Excessive Vitamins and Minerals in Food Put Millions of Children at Risk," Environmental Working Group press release, June 24, 2014, https://www.ewg.org/release/excessive-vitamins-and-minerals-food-put-millions-children-risk#.Wk8A-YSYe_U.
31. Shi-Sheng Zhou and Yiming Zhou, "Excess Vitamin Intake: An Unrecognized Risk Factor for Obesity," *World Journal of Diabetes* 5, no. 1 (February 2014): 1–13; Jill Reedy et al., "Dietary Sources of Energy, Solid Fats, and Added Sugars Among Children and Adolescents in the United States," *Journal of the American Dietetic Association* 110, no. 10 (October 2010): 1477–84.
32. P. Mersereau et al., "Spina Bifida and Anencephaly Before and After Folic Acid Mandate—United States, 1995–1996 and 1999–2000," *Morbidity and Mortality Weekly Report* 53, no. 17 (May 7, 2004): 362–65, http://www.cdc.gov/mmwr/preview/mmwrhtml/mm5317a3.htm; Tom D. Thacher et al., "Increasing Incidence of Nutritional Rickets: A Population-Based Study in Olmsted County, Minnesota," *Mayo Clinic Proceedings* 88, no. 2 (February 2013): 176–83; Sally C. Davies, *Annual Report of the Chief Medical Officer 2012: Our Children Deserve Better: Prevention Pays*, U.K. Department of Health, October 2013, https://www.gov.uk/government/uploads/system/uploads/attachment_data/file/255237/2901304_CMO_complete_low_res_accessible.pdf.

CHAPTER FOUR: THE INDUSTRIAL FOOD SETUP

1. Nielsen, "Oh Baby! Baby Food and Formula Sales Will Reach Nearly $30 Billion Around the World," Nielson Global Baby Care Survey press release, August, 25, 2015, http://www.nielsen.com/us/en/press-room/2015/oh-baby-global-baby-food-and-formula-sales-ill-reach-nearly-35-billion.html.
2. Kristin Wartman, "Bad Eating Habits Start in the Womb," *The New York Times*, December 1, 2013, http://www.nytimes.com/2013/12/02/opinion/bad-eating-habits-start-in-the-womb.html.
3. Julie A. Mennella, interview by author, November 5, 2013, Philadelphia, Pennsylvania.
4. Ibid.

5. Julie A. Mennella and Nuala K. Bobowski, "The Sweetness and Bitterness of Childhood: Insights from Basic Research on Taste Preferences," *Physiological Behavior* 152 (December 1, 2015): 502–7; Mennella interview.
6. Malia Wollan, "Rise and Shine: What Kids Around the World Eat for Breakfast," *The New York Times Magazine*, October 8, 2014, https://www.nytimes.com/interactive/2014/10/08/magazine/eaters-all-over.html.
7. Bettina T. Cornwell and A. R. McAlister, "Alternative Thinking About Starting Points of Obesity. Development of Child Taste Preferences," *Appetite* 56, no. 2 (April 2011): 428–39.
8. Bettina T. Cornwell, interview by author, December 21, 2015, Eugene, Oregon.
9. Kate Northstone, Pauline M. Emmet, and the ALSPAC Study Team, "Multivariate Analysis of Diet in Children at Four and Seven Years of Age and Associations with Socio-Demographic Characteristics," *European Journal of Clinical Nutrition* 59, no. 6 (June 2005): 751–60.
10. Mennella interview; Cornwell interview.
11. Cornwell interview.
12. Anna R. McAlister and T. Bettina Cornwell, "Collectible Toys as Marketing Tools: Understanding Preschool Children's Responses to Foods Paired with Premiums," *Journal of Public Policy & Marketing* 31, no. 2 (Fall 2012): 195–205.
13. Marsha Walker, email to author, November 4, 2013.
14. Kathy Dettwyler, interview by author, November 4, 2013, Newark, Delaware.
15. Angela Garbes, "The More I Learn About Breast Milk, the More Amazed I Am," *The Stranger*, August 26, 2015, http://www.thestranger.com/features/feature/2015/08/26/22755273/the-more-i-learn-about-breast-milk-the-more-amazed-i-am; Olivia Ballard et al., "Human Milk Composition: Nutrients and Bioactive Factors," *Pediatric Clinics of North America* 60, no. 1 (February 2013): 49–74.
16. Public Citizen, "Fact Sheet: Infant Formula Marketing in Healthcare Facilities," October 2013, https://www.citizen.org/our-work/health-and-safety/infant-formula-marketing-healthcare-facilities. For full

report, https://www.citizen.org/sites/default/files/best-hospitals-end-infant-formula-marketing-to-support-breastfeeding-report.pdf.

17. Sandra Gordon, "What Should We Feed Baby?" Parent's Choice Formula, no date, http://www.parentschoiceformula.com/articles/Infant-Formula-Timeline-What-Should-we-Feed-Baby.aspx (accessed November 2, 2015).
18. Amy Bentley, "Booming Baby Food: Infant Food and Feeding in Post–World War II America," *Michigan Historical Review* 32, no. 2 (Fall 2006): 63–87.
19. U.S. Centers for Disease Control and Prevention, "Breastfeeding Report Card 2016," August 2016, https://www.cdc.gov/breastfeeding/pdf/2016breastfeedingreportcard.pdf; U.S. Centers for Disease Control and Prevention, "Breastfeeding: Frequently Asked Questions (FAQs)," June 16, 2015, https://www.cdc.gov/breastfeeding/faq/; Office on Women's Health, U.S. Department of Health and Human Services, "Breastfeeding," May 3, 2017, https://www.womenshealth.gov/Breastfeeding/ (accessed August 8, 2017).
20. Christopher G. Owen et al., "Effect of Infant Feeding on the Risk of Obesity Across the Life Course: A Quantitative Review of Published Evidence," *Pediatrics* 115, no. 5 (2005): 1367–77; M. Jane Heinig et al., "Energy and Protein Intakes of Breast-Fed and Formula-Fed Infants During the First Year of Life and Their Association with Growth Velocity: The DARLING Study," *American Journal of Clinical Nutrition* 58, no. 2 (1993): 152–61.
21. Atul Singhal et al., "Early Nutrition and Leptin Concentrations in Later Life," *American Journal of Clinical Nutrition* 75, no. 6 (2002): 993–99; Joanna Overberg et al., "Differences in Taste Sensitivity Between Obese and Non-Obese Children and Adolescents," *Archives of Disease in Childhood* 97, no. 12 (September 2012): 1048–52.
22. M. Yanina Pepino and Julie A. Mennella, "Sucrose-Induced Analgesia Is Related to Sweet Preferences in Children but Not Adults," *Pain*, 119 (1–3), (November 18, 2005): 210–18.
23. Mennella, interview.
24. Michael Chevrot et al., "Obesity Interferes with the Orosensory Detection of Long-Chain Fatty Acids in Humans," *The American Journal of Clinical Nutrition* 99, no. 5 (May 2014): 975–83.
25. Wartman, "Bad Eating Habits Start in the Womb."

26. Courtney Jung, *Lactivism: How Feminists and Fundamentalists, Hippies and Yuppies, and Physicians and Politicians Made Breastfeeding Big Business and Bad Policy* (New York: Basic Books, 2015).

CHAPTER FIVE: LIVING IN A MICROBIAL WORLD

1. J. Bruce German, interview by author, March 2, 2017, Davis, California.
2. German, interview.
3. U.S. Centers for Disease Control and Prevention, "CDC: 1 in 3 Antibiotic Prescriptions Unnecessary," press release, May 3, 2016, https://www.cdc.gov/media/releases/2016/p0503-unnecessary-prescriptions.html.
4. Mark A. Underwood et al., "*Bifidobacterium Longum* Subspecies *Infantis*: Champion Colonizer of the Infant Gut," *Pediatric Research* 77 (2015): 229–35.
5. German, interview.
6. German, interview.
7. Tommi Vatanen et al., "Variation in Microbiome LPS Immunogenicity Contributes to Autoimmunity in Humans," *Cell* 165, no. 4 (May 2016): 842–53.
8. German, interview.
9. J. Bruce German, series of emails with author, June 10, 11, 2016, and August 26, 2016.
10. Vatanen et al., "Variation in Microbiome LPS Immunogenicity," 842–53.
11. Quang Nguyen et al., "The Impact of the Gut Microbiota on Humoral Immunity to Pathogens and Vaccination in Early Infancy," *PLoS Pathogens* 12, no. 12 (December 2016), https://doi.org/10.1371/journal.ppat.1005997.
12. German, interview.
13. Rebecca L. Siegel, "Colorectal Cancer Incidence Patterns in the United States, 1974–2013," *Journal of the National Cancer Institute* 109, no. 8 (February 28, 2017), https://doi.org/10.1093/jnci/djw322.
14. Cochrane reviewed the existing literature and concluded that evidence from well-designed and -conducted trials is insufficient to recommend

antibiotic use in labor to reduce the risk of Group B strep infection in newborns. Arne Ohlsson and Vibhuti S. Shah, "Intrapartum Antibiotics for Known Maternal Group B Streptococcal Colonization," Cochrane.org, June 10, 2014, http://www.cochrane.org/CD007467/PREG_intrapartum-antibiotics-known-maternal-group-b-streptococcal-colonization.

15. Martin Blaser, "The Way You're Born Can Mess with the Microbes You Need to Survive," *Wired Magazine*, April 13, 2014, https://www.wired.com/2014/04/missing-microbes-antibiotic-resistance-birth/.
16. Noel T. Mueller et al., "The Infant Microbiome Development: Mom Matters," *Trends in Molecular Medicine* 21, no. 2 (February 2015): 109–17.
17. Ibid.
18. J. Bruce German, interview by author, June 5, 2016, Davis, California.
19. Martin J. Blaser, "Antibiotic Overuse: Stop the Killing of Beneficial Bacteria," *Nature* 476, no. 7361 (August 2011): 393–94.
20. Paul Forsythe et al., "Vagal Pathways for Microbiome-Brain-Gut Axis Communication," *Advances in Experimental Medicine and Biology* 817 (June 9, 2014): 115–33.
21. J. Bruce German, interview with author, May 27, 2016, Davis, California.
22. Dr. Richard Frye, interview by author, May 12, 2016, Fayetteville, Arkansas.
23. Ibid.
24. Ibid.
25. Ibid.
26. J. Bruce German, interview with author, March 2, 2017, Davis, California.
27. Erica D. Sonnenburg and Justin L. Sonnenburg, "Starving our Microbial Self: The Deleterious Consequences of a Diet Deficient in Microbiota-Accessible Carbohydrates," *Cell Metabolism* 20, issue 5 (November 4, 2014): 779–86.
28. Erica Sonnenburg, interview with author, January 15, 2016, Stanford, California.
29. Erica D. Sonnenburg et al., "Diet-Induced Extinction in the Gut Microbiota Compounds Over Generations," *Nature* 529, no. 7585 (January 14, 2016): 212–15.
30. Sonnenburg, interview.

31. Rebecca L. Siegel et al., "Colorectal Cancer Incidence Patterns in the United States, 1974–2013," *Journal of the National Cancer Institute* 109, no. 8 (February 28, 2017), https://academic.oup.com/jnci/article-lookup/doi/10.1093/jnci/djw322.
32. Alexandra L. McOrist et al., "Fecal Butyrate Levels Vary Widely Among Individuals but Are Usually Increased by a Diet High in Resistant Starch," *The Journal of Nutrition* 141, no. 5 (May 2011): 883–89.
33. Gijs den Besten et al., "Gut-Derived Short-Chain Fatty Acids Are Vividly Assimilated into Host Carbohydrates and Lipids," *American Journal of Physiology Gastrointestinal and Liver Physiology* 305, no. 12 (December 2013): G900–910.
34. Benoit Chassaing et al., "Dietary Emulsifiers Impact the Mouse Gut Microbiota Promoting Colitis and Metabolic Syndrome," *Nature* 519 (March 5, 2015): 92–96.
35. Andrew Gewirtz, interview by author, January 22, 2016, Atlanta, Georgia.
36. Ibid.
37. Jotham Suez et al., "Artificial Sweeteners Induce Glucose Intolerance by Altering the Gut Microbiota," *Nature* 514 (October 2014): 181–86.
38. Eran Segal, interview by author, January 28, 2016, Rehovot, Israel.
39. Alistair B. A. Boxall et al., "Uptake of Veterinary Medicines from Soils into Plants," *Journal of Agriculture and Food Chemistry* 54, no. 6 (2006): 2288–97; James Hamblin, "Are Antibiotics Making People Larger?" *The Atlantic*, December 21, 2015, https://www.theatlantic.com/health/archive/2015/12/obesity-antibiotics-microbiome/421344/.
40. Martin J. Blaser, *Missing Microbes* (New York: Henry Holt, 2014), 85; Shinwoo Yang and Ken Carlson, "Evolution of Antibiotic Occurrence in a River Through Pristine, Urban and Agricultural Landscapes," *Water Research* 37, no. 19 (November 2003): 4645–56; Hamblin, "Are Antibiotics Making People Larger?"; Anthony Kappell et al., "Detection of Multi-Drug Resistant Escherichia Coli in the Urban Waterways of Milwaukee, WI," *Frontiers in Microbiology* 6, no. 336 (April 2015), https://www.ncbi.nlm.nih.gov/pubmed/25972844.

41. Ransdell Pierson and Bill Berkrot, "U.S. Sees First Case of Bacteria Resistant to Last-Resort Antibiotic," Reuters, May 27, 2016; U.S. Food and Drug Administration, *2015 Summary Report on Antimicrobials Sold or Distributed for Use in Food-Producing Animals* (Laurel, MD: Center for Veterinary Medicine, December 2016), https://www.fda.gov/downloads/ForIndustry/UserFees/AnimalDrugUserFeeActADUFA/UCM534243.pdf.
42. Laura M. Cox et al., "Altering the Intestinal Microbiota During a Critical Developmental Window Has Lasting Metabolic Consequences," *Cell* 158, no. 4 (August 2014): 705–21; Blaser, *Missing Microbes.*
43. Charles M. Benbrook, "Trends in Glyphosate Herbicide Use in the United States and Globally," *Environmental Sciences Europe* 28, no. 3 (January 11, 2016), https://enveurope.springeropen.com/articles/10.1186/s12302-016-0070-0.
44. Monika Kruger et al., "Glyphosate Suppresses the Antagonistic Effect of *Enterococcus ssp.* on *Clostridium botulinum*," *Anaerobe* 20 (April 2013): 74–78.
45. Benoit Chassaing et al., "Lack of Soluble Fiber Drives Diet-Induced Adiposity in Mice," *American Journal of Physiology* 309, no. 7 (October 2015): G528–41.
46. Gewirtz, interview.
47. Tommi Vantanen et al., "Variation in Microbiome LPS Immunogenicity Contributes to Autoimmunity in Human," *Cell* 165, no. 4 (May 2016): 842–53.
48. Sonnenburg, interview.

CHAPTER SIX: "SO MANY TIPS OF SO MANY ICEBERGS"

1. Charles Darwin to Moritz Wagner, October 13, 1876, in *Life and Letters of Charles Darwin*, vol. 2, http://charles-darwin.classic-literature.co.uk/the-life-and-letters-of-charles-darwin-volume-ii/ebook-page-156.asp.
2. Christopher J. L. Murray et al., "Disability-Adjusted Life Years (DALYs) for 291 Diseases and Injuries in 21 Regions, 1990–2010: A Systematic

Analysis for the Global Burden of Disease Study 2010," *The Lancet* 380, no. 9859 (December 15, 2012): 2197–223.

3. Åke Bergman et al., eds., *State of the Science of Endocrine Disrupting Chemicals 2012* (Geneva: United Nations Environment Programme and World Health Organization, 2013), http://www.who.int/ceh/publications/endocrine/en/; Breast Cancer Prevention Partners, "Breast Cancer Statistics," https://www.bcpp.org/resource/breast-cancer-statistics/ (accessed April 3, 2017); American Cancer Society, "How Common Is Breast Cancer?," September 27, 2017, https://www.cancer.org/cancer/breast-cancer/about/how-common-is-breast-cancer.html.
4. M. B. Macon and S. E. Fenton, "Endocrine Disruptors and the Breast: Early Life Effects and Later Life Disease," *Journal of Mammary Gland Biology and Neoplasia* 18, no. 1 (March 2013): 43–61; Robert Martin, "Toxic Time Bombs," *The Scientist*, September 25, 2017, http://www.the-scientist.com/?articles.view/articleNo/50485/title/Opinion—Toxic-Time-Bombs/; U.S. Centers for Disease Control and Prevention, "Dichlorodiphenyltrichloroethane (DDT)," November 2009, https://www.cdc.gov/biomonitoring/pdf/ddt_factsheet.pdf.
5. R. Thomas Zoeller et al., "Endocrine-Disrupting Chemicals and Public Health Protection: A Statement of Principles from the Endocrine Society," *Endocrinology* 153, no. 9 (June 4, 2012): 4097–4110; Andrea C. Gore et al., "EDC-2: The Endocrine Society's Second Scientific Statement on Endocrine-Disrupting Chemicals," *Endocrine Reviews* 36, no. 6 (December 2015): E1–E150; Andrea C. Gore et al., "Executive Summary to EDC-2: The Endocrine Society's Second Scientific Statement on Endocrine-Disrupting Chemicals," *Endocrine Reviews* 36, no. 6 (December 2015): 593–602.
6. Jerrold J. Heindel et al., "Metabolism Disrupting Chemicals and Metabolic Disorders," *Reproductive Toxicology* 68 (March 2017): 3–33; National Institute of Environmental Health Sciences, "Endocrine Disruptors," no date, https://www.niehs.nih.gov/health/topics/agents/endocrine/ (March 9, 2017).
7. Bruce Blumberg, interview by author, March 30, 2016, Irvine, California; R. Thomas Zoeller, interview by author, November 20, 2017, Amherst, Massachusetts.

8. Ibid.
9. Douglas Main, "BPA Is Fine, If You Ignore Most Studies About It," *Newsweek*, March 15, 2015, http://www.newsweek.com/2015/03/13/bpa-fine-if-you-ignore-most-studies-about-it-311203.html.
10. Frederick vom Saal, interview by author, April 16, 2015, Columbia, Missouri.
11. Laura Vandenberg, interview by author, April 21, 2016, Amherst, Massachusetts.
12. Ibid.
13. Vom Saal, interview; Brittany M. Angle et al., "Metabolic Disruption in Male Mice Due to Fetal Exposure to Low but Not High Doses of Bisphenol A (BPA): Evidence for Effects on Body Weight, Food Intake, Adipocytes, Leptin, Adiponectin, Insulin and Glucose Regulation," *Reproductive Toxicology* 42 (July 2013): 256–68.
14. Fabiana Ariemma et al., "Low-Dose Bisphenol-A Impairs Adipogenesis and Generates Dysfunctional 3T3-L1 Adipocytes," *PLoS One*, March 4, 2016, http://journals.plos.org/plosone/article?id=10.1371/journal.pone.0150762; vom Saal, interview.
15. A. Ziv-Gal et al., "The Effects of In Utero Bisphenol A Exposure on Reproductive Capacity in Several Generations of Mice," *Toxicology and Applied Pharmacology* 284, no. 3 (May 1, 2015): 354–62; vom Saal, interview.
16. F. Grün and B. Blumberg, "Environmental Obesogens: Organotins and Endocrine Disruption via Nuclear Receptor Signaling," *Endocrinology* 147, no. 6 (2006): S50–S55.
17. Kristin Wartman, "What's Really Making Us Fat?" *The Atlantic*, March 8, 2012, https://www.theatlantic.com/health/archive/2012/03/whats-really-making-us-fat/254087/.
18. Raquel Chamorro-Garcia et al., "Transgenerational Inheritance of Increased Fat Depot Size, Stem Cell Reprogramming, and Hepatic Steatosis Elicited by Prenatal Exposure to the Obesogen Tributyltin in Mice," *Environmental Health Perspectives* 121 (March 1, 2013), https://ehp.niehs.nih.gov/1205701/.
19. Dr. Sarah Wartman, interview by author, November 8, 2016, Los Angeles, California; Blumberg, interview.
20. Andrea Gore, interview by author, April 2, 2017, Austin, Texas.

21. Matthew D. Anway et al., "Epigenetic Transgenerational Actions of Endocrine Disruptors and Male Fertility," *Science* 308, no. 5727 (June 3, 2005): 1466–69.
22. Gore, interview; David Crews et al., "Transgenerational Epigenetic Imprints on Mate Preference," *Proceedings of the National Academy of Sciences* 104, no. 14 (March 26, 2007): 5942–46.
23. Vom Saal, interview.
24. Gore, interview.
25. Ibid.
26. Lawrence Wright, "Silent Sperm," *The New Yorker*, January 15, 1996; Liborio Stuppia et al., "Epigenetics and Male Reproduction: The Consequences of Paternal Lifestyle on Fertility, Embryo Development, and Children Lifetime Health," *Clinical Epigenetics* 7, no. 120 (November 2015), https://www.ncbi.nlm.nih.gov/pubmed/26566402; Hagai Levine et al., "Temporta Trends in Sperm Count: A Systemic Review and Meta-Regression Analysis," *Human Reproduction Update* 23, no. 6 (November 1, 2017): 646–59.
27. Chris Barratt, "Most Men in the US and Europe Could Be Infertile by 2060, According to New Study," *Quartz*, July 28, 2017, https://qz.com/1040302/most-men-in-the-us-and-europe-could-be-infertile-by-2060-according-to-a-new-study/; Gore, interview.
28. Gore, interview; "Dioxins and Their Effects on Human Health," World Health Organization fact sheet, October 2016, http://www.who.int/mediacentre/factsheets/fs225/en/.
29. "Dioxins and Their Effects on Human Health."
30. Marie-Monique Robin, *The World According to Monsanto: Pollution, Corruption, and the Control of the World's Food Supply* (New York: New Press, 2010), 13.
31. Sarah M. Dickerson, Stephanie L. Cunningham, and Andrea C. Gore, "Prenatal PCBs Disrupt Early Neuroendocrine Development of the Rat Hypothalamus," *Toxicology and Applied Pharmacology* 252, no. 1 (April 1, 2011): 36–46.
32. Rebecca M. Steinberg, Thomas E. Juenger, and Andrea C. Gore, "The Effects of Prenatal PCBs on Adult Female Paced Mating Reproductive Behaviors in Rats," *Hormones and Behavior* 51, no. 3 (March 2007): 364–72.

33. Gore, interview.
34. Margaret Bell, Bethany G. Hart, and Andrea C. Gore, *Molecular and Cellular Endocrinology* 420 (January 2016):125–37.
35. Margaret R. Bell, Bethany G. Hart, and Andrea C. Gore, "Two-Hit Exposure to Polychlorinated Biphenyls at Gestational and Juvenile Life Stages: 2. Sex-Specific Neuromolecular Effects in the Brain," *Molecular and Cellular Endocrinology* 420 (January 15, 2016): 125–37.
36. Gore, interview.
37. Blumberg, interview; Zoeller, interview.
38. Zoeller, interview.
39. Gore, interview.
40. Andrea Gore et al., "Early Life Exposure to Endocrine-Disrupting Chemicals Causes Lifelong Molecular Reprogramming of the Hypothalamus and Premature Reproductive Aging," *Molecular Endocrinology* 25, no. 12 (December 2011): 2157–68; Gore, interview.
41. Ruthann M. Giusti, Kumiko Iwamoto, and Elizabeth E. Hatch, "Diethylstilbestrol Revisited: A Review of the Long-Term Health Effects," *Annals of Internal Medicine* 122, no. 10 (May 15, 1995): 778–88.
42. Janneke Verloop, Flora E. van Leeuwen, Theo J. M. Helmerhorst et al., "Cancer Risk in DES Daughters," *Cancer Causes & Control* 21, no. 7 (July 2010): 999–1007.
43. Julie R. Palmer, Arthur L. Herbst, Kenneth L. Noller et al., "Urogenital Abnormalities in Men Exposed to Diethylstilbestrol In Utero: A Cohort Study," *Environmental Health* 8, no. 37 (2009), https://doi.org/10.1186/1476-069X-8-37.
44. National Cancer Institute, "Diethylstilbestrol (DES) and Cancer," October 5, 2011, https://www.cancer.gov/about-cancer/causes-prevention/risk/hormones/des-fact-sheet#q3; The DES Follow-up Study, https://www.desfollowupstudy.org/index.asp (accessed January 22, 2018).
45. Vandenberg, interview.
46. Ibid.
47. Gore, interview, April 5, 2016, Austin, Texas.
48. Blumberg, email to author, March 19, 2016.
49. Blumberg, interview; Gore, interview.
50. Charles M. Benbrook, "Trends in Glyphosate Herbicide Use in the United States and Globally," *Environmental Sciences Europe* 28, no. 3

(2016), http://enveurope.springeropen.com/articles/10.1186/s12302-016-0070-0; John Peterson Myers et al., "Concerns over Use of Glyphosate-Based Herbicides and Risks Associated with Exposures: A Consensus Statement," *Environmental Health* 15, no. 19 (February 17, 2016), https://doi.org/10.1186/s12940-016-0117-0.

51. Zoë Schlanger, "The FDA Will Begin Testing Food for Glyphosate, the Most Heavily Used Farm Chemical Ever," *Newsweek*, February 19, 2016, http://www.newsweek.com/fda-will-begin-testing-food-glyphosate-most-heavily-used-farm-chemical-ever-428790 (accessed June 5, 2017); Carey Gillam, "FDA Suspends Testing for Glyphosate Residues in Food," November 11, 2016, http://www.huffingtonpost.com/carey-gillam/fda-suspends-glyphosate-r_b_12913458.html; Benbrook, "Trends in Glyphosate Herbicide Use."

CHAPTER SEVEN: OUR "SAFE" EXPOSURE TO TOXIC CHEMICALS

1. Roni Caryn Rabin, "The Chemicals in Your Mac and Cheese," *The New York Times*, July 12, 2017; Stephanie Strom, "Traces of Controversial Herbicide Are Found in Ben & Jerry's Ice Cream," *The New York Times*, July 25, 2017; Ben and Jerry's said it will stop sourcing ingredients made with crops that use glyphosate as a drying agent by 2020, https://www.theguardian.com/environment/2017/oct/09/ben-jerrys-to-launch-glyphosate-free-ice-cream-after-tests-find-traces-of-weedkiller.
2. Thomas G. Neltner et al., "Data Gaps in Toxicity Testing of Chemicals Allowed in Food in the United States," *Reproductive Toxicology* 42 (December 2013): 85–94.
3. Bruce Blumberg, interview by author, April 13, 2016, Irvine, California.
4. R. Thomas Zoeller, interview by author, November 20, 2017, Amherst, Massachusetts.
5. Maricel Maffini, interview by author, July 25, 2016, Germantown, Maryland.
6. Blumberg, interview.
7. Sabinsa Corporation, "GRAS Notice Number 474: Piperine Derived from the Fruits of *Piper nigrum* L (Black Pepper) or *P. longum* (Long Pepper)," submitted to the U.S. Food and Drug Administration, May 20, 2013, https://www.fda.gov/downloads/Food/IngredientsPackagingLabeling

/GRAS/NoticeInventory/ucm362936.pdf (accessed September 12, 2017). GRAS stands for generally regarded as safe.

8. Zoeller, interview.
9. Maricel Maffini interview, January 13, 2017, Germantown, Maryland.
10. Tom Neltner and Maricel Maffini, "Generally Recognized as Secret: Chemicals Added to Foods in the United States," *Natural Resources Defense Council Report*, April 2014, https://www.nrdc.org/sites/default/files/safety-loophole-for-chemicals-in-food-report.pdf.
11. Maffini, interview.
12. Neltner and Maffini, "Generally Recognized as Secret."
13. Maffini, interview.
14. Neltner and Maffini, "Generally Recognized as Secret."
15. Neltner et al., "Data Gaps in Toxicity Testing"; Maffini, interview; Kimberly Kindy, "Food Additives on the Rise as FDA Scrutiny Wanes," *The Washington Post*, August 17, 2014; Megan McSeveney, email to author, April 20, 2017.
16. Maffini, interview.
17. Erin Quinn and Chris Young, "Why the FDA Doesn't Really Know What's in Your Food," The Center for Public Integrity, April 14, 2015, https://www.publicintegrity.org/2015/04/14/17112/why-fda-doesnt-really-know-whats-your-food.
18. "3-Ex Officials of Major Laboratory Convicted of Falsifying Drug Tests," *The New York Times*, October 22, 1983, http://www.nytimes.com/1983/10/22/us/3-ex-officials-of-major-laboratory-convicted-of-falsifying-drug-tests.html.
19. Laura Vandenberg, interview by author, April 21, 2016, Amherst, Massachusetts.
20. Maffini, interview.
21. Ibid.
22. U.S. Food and Drug Administration, "Bisphenol A (BPA): Use in Food Contact Application," press release, January 2010 (last updated November 2014), https://www.fda.gov/newsevents/publichealthfocus/ucm064437.htm (accessed September 3, 2017); Sandra Laville and Matthew Taylor, "A Million Bottles a Minute: World's Plastic Binge 'as Dangerous as Climate Change,'" *The Guardian*, June 28, 2017, https://www.theguardian

.com/environment/2017/jun/28/a-million-a-minute-worlds-plastic-bottle-binge-as-dangerous-as-climate-change.

23. John Peterson Myers et al., "Abstract: Why Public Health Agencies Cannot Depend on Good Laboratory Practices as a Criterion for Selecting Data: The Case of Bisphenol A," *Environmental Health Perspectives* 117, no. 3 (March 2009): 309–15.
24. Maffini, interview.
25. Maffini, email with author, August 22, 2017.
26. Julia A. Taylor et al., "Similarity of Bisphenol A Pharmacokinetics in Rhesus Monkeys and Mice: Relevance for Human Exposure," *Environmental Health Perspectives* 110, no. 4 (April 2011): 424–30.
27. Daniel R. Doerge et al., "Pharmacokinetics of Bisphenol A in Neonatal and Adult Rhesus Monkeys," *Toxicology and Applied Pharmacology* 248, no. 1 (October 2010): 1–11.
28. Frederick S. vom Saal, interview by author, April 4, 2015, Columbia, Missouri.
29. Letter to the director of the National Center for Toxicology Research, Dr. William Slikker, and the associate director of the National Toxicology Program, Dr. John R. Bucher (both organizations are part of the FDA) from Patricia Hunt, professor of molecular biosciences at Washington State University, et al., October 1, 2010.
30. Daniel Doerge, email to author, December 18, 2017.
31. Zoeller, interview.
32. Ibid.
33. Elizabeth Flock, "Monsanto Petition Tells Obama: 'Cease FDA Ties to Monsanto,'" *The Washington Post*, January 30, 2012, https://www.washingtonpost.com/blogs/blogpost/post/monsanto-petition-tells-obama-cease-fda-ties-to-monsanto/2012/01/30/gIQAA9dZcQ_blog.html; Tom Philpott, "Monsanto's Man Taylor Returns to FDA in Food-Czar Role," *Grist*, July 9, 2009, http://grist.org/article/2009-07-08-monsanto-fda-taylor/.
34. "Dr. Mitchell Cheeseman," *Steptoe & Johnson LLP* website, https://www.steptoe.com/professionals-Mitchell_Cheeseman.html (accessed January 8, 2018); Matthew Huisman, "Two Former FDA Officials Join Steptoe & Johnson's D.C. Office," *The BLT: The Blog of LegalTimes*, November 29,

2011, http://legaltimes.typepad.com/blt/2011/11/two-former-fda-officials-join-steptoe-johnsons-dc-office.html (accessed January 8, 2018); Megan Stride, "Steptoe Snags 2 FDA Food Safety Pros For DC Office," *Law360*, November 28, 2011, https://www.law360.com/articles/288688/steptoe-snags-2-fda-food-safety-pros-for-dc-office.

35. Vom Saal, interview; Zoeller, interview.
36. Associated Press, "EPA Chief Met with Dow Chemical CEO Before Deciding Not to Ban Toxic Pesticide," June 30, 2017, http://www.latimes.com/business/la-fi-epa-pesticide-dow-20170627-story.html; Eric Lipton, "Why Has the E.P.A. Shifted on Toxic Chemicals? An Industry Insider Helps Call the Shots," *The New York Times*, October 21, 2017, https://www.nytimes.com/2017/10/21/us/trump-epa-chemicals-regulations.html.
37. Kristin Wartman Lawless, "No One Knows How Much Herbicide Is in Your Breakfast," *Vice*, May 2016, https://www.vice.com/en_us/article/no-one-knows-how-much-herbicide-is-in-your-breakfast.
38. Blumberg, interview; Zoeller, interview.

CHAPTER EIGHT: FOOD CHOICE

1. "President Obama," *Real Time with Bill Maher*, HBO, November 4, 2016.
2. Kate Taylor, "These 10 Companies Control Everything You Buy," *Business Insider*, September 28, 2016, http://www.businessinsider.com/10-companies-control-the-food-industry-2016-9.
3. Bill McKibben, *Deep Economy: The Wealth of Communities and the Durable Future* (New York: Henry Holt, 2007), 52–53.
4. Judy S. LaKinda and Daniel Q. Naiman, "Daily Intake of Bisphenol A and Potential Sources of Exposure: 2005–2006 National Health and Nutrition Examination Survey," *Journal of Exposure Science and Environmental Epidemiology* 21 (2011): 272–79; Antonia Calafat, X Ye, et al., "Exposure of the U.S. Population to Bisphenol A and 4-Tertiary-Octylphenol," *Environmental Health Perspectives* 116, no. 1 (2008): 39–44; Tom Philpott, "The Rich Are Eating Richer, the Poor Are Eating Poorer," *Mother Jones*, September 11, 2014, http://www.motherjones.com/food/2014/09/food-inequality/.
5. Eric Holt-Giménez and Breeze Harper, "Food—Systems—Racism:

From Mistreatment to Transformation," *Food First*, inter-spring 2016, https://foodfirst.org/wp-content/uploads/2016/03/DR1Final.pdf.

6. Amanda Bruce et al., "Branding and a Child's Brain: An fMRI Study of Neural Responses to Logos," *Social, Cognitive, and Affective Neuroscience* 9, no. 1 (September 2012): 118–22.
7. Jennifer Harris et al., "Fast Food FACTS 2013 Report," Yale Rudd Center for Food Policy and Obesity, November 2013, https://www.rwjf.org/content/dam/farm/reports/reports/2013/rwjf408549; Frances Fleming-Milici and J. L. Harris, "Television Food Advertising Viewed by Preschoolers, Children and Adolescents: Contributors to Differences in Exposure for Black and White Youth in the United States," *Pediatric Obesity* (November 2016), http://dx.doi.org/10.1111/ijpo.12203.
8. Richard Manning, *Against the Grain: How Agriculture Has Hijacked Civilization* (New York: North Point Press, 2004), 181; Cynthia L. Ogden et al., "Obesity and Socioeconomic Status in Adults: United States, 2005–2008," *National Center for Health Statistics Data Brief* 50 (December 2010), https://www.cdc.gov/nchs/data/databriefs/db50.pdf; Margarita Teran-Garcia et al., "Impact of Acculturation on Dietary Habits of Latina Immigrants," *The Journal of the Federation of American Societies for Experimental Biology* 25, no. 1 (April 2011), http://www.fasebj.org/content/25/1_Supplement/974.12.short.
9. Kristin Wartman, "Not Lovin' It: McDonald's Wages Class War in New Ads," *Huffington Post*, January 25, 2015, https://www.huffingtonpost.com/kristin-wartman/not-lovin-it-mcdonalds-wa_b_6531918.html.
10. Valerie Strauss, "Palin: Parents Should Decide What Kids Eat in School," *The Washington Post*, November 12, 2010; Kristin Wartman, "Bloomberg vs. Beyoncé: The Real Dilemma with NYC's Soda Ban," *Civil Eats*, June 13, 2013, https://civileats.com/2013/06/13/bloomberg-vs-beyonce-the-real-dilemma-with-nycs-soda-ban/.
11. Wartman, "Bloomberg vs. Beyoncé."
12. Manning, *Against the Grain*, 178.
13. Felipe Fernández-Armesto, *Near a Thousand Tables: A History of Food* (New York: Free Press, 2002), 123.
14. Kristin Wartman, "The American Fast Food Syndrome," *Civil Eats*, January 13, 2011, https://civileats.com/2011/01/13/the-american-fast-food-syndrome/.

15. Michael Pollan, *In Defense of Food: An Eater's Manifesto* (New York: Penguin, 2008).
16. "Alice Waters' Crusade for Better Food," *60 Minutes*, CBS, March 15, 2009, http://www.cbsnews.com/news/alice-waters-crusade-for-better-food/.
17. Selma James, *The Power of Women and the Subversion of the Community*," in Selma James, *Sex, Race, and Class—The Perspective of Winning: A Selection of Writings, 1952–2011*, 3rd ed. (Oakland, CA: PM Press), 57; emphasis in the original.
18. Mark Bittman, "Is Junk Food Really Cheaper?" *The New York Times*, September 24, 2011; Tracie McMillan, *The American Way of Eating* (New York: Scribner, 2012), 115.
19. Robert Reich, "The 'Paid What You're Worth' Myth," *Huffington Post*, March 14, 2014, https://www.huffingtonpost.com/robert-reich/paid-what-youre-worth_b_4964290.html (accessed January 4, 2016).
20. Wendy Wang et al., "Breadwinner Moms," *Pew Research Center Social and Demographic Trends*, May 29, 2013, http://www.pewsocialtrends.org/2013/05/29/breadwinner-moms/.

CHAPTER NINE: HANDMAIDENS OF INDUSTRY

1. Cria G. Perrine et al., "Baby-Friendly Hospital Practices and Meeting Exclusive Breastfeeding Intention," *Pediatrics* 130, no. 1 (May 2012): 54–60; Gretchen Livingston, "Among 41 Nations, U.S. Is the Outlier When It Comes to Paid Parental Leave," *Fact Tank*, September 26, 2016, http://www.pewresearch.org/fact-tank/2016/09/26/u-s-lacks-mandated-paid-parental-leave/.
2. Report from the Organization for Economic Co-operation and Development, 2010, https://data.oecd.org.
3. Michael Pollan, *Cooked: A Natural History of Transformation* (New York: Penguin, 2013), 10.
4. Glenna Matthews, *"Just a Housewife": The Rise & Fall of Domesticity in America* (New York: Oxford University Press, 1987), 35.
5. Maxine Margolis, *Mothers and Such: Views of American Women and Why They Changed* (Berkeley: University of California Press, 1984), 28.
6. Matthews, *"Just a Housewife,"* 112.
7. Ibid., 179.

8. Ibid., 104; John L. Hess and Karen Hess, *The Taste of America* (Urbana and Chicago: University of Illinois Press, 2000), 105.
9. Matthews, *"Just a Housewife,"* 105–6.
10. Laura Shapiro, *Perfection Salad: Women and Cooking at the Turn of the Century* (Berkeley: University of California Press, 2009), 209, 212.
11. John Kenneth Galbraith, *Economics and the Public Purpose* (Boston: Houghton Mifflin, 1973), 33; Jenna Goudreau, "Why Stay-at-Home-Moms Should Earn a $115,000 Salary," *Forbes*, May 2, 2011, https://www.forbes.com/sites/jennagoudreau/2011/05/02/why-stay-at-home-moms-should-earn-a-115000-salary/.
12. Goudreau, "Why Stay-At-Home-Moms Should Earn a $115,000 Salary."
13. Selma James, *The Power of Women and the Subversion of the Community,*" in Selma James, *Sex, Race, and Class—The Perspective of Winning: A Selection of Writings, 1952–2011*, 3rd ed. (Oakland, CA: PM Press), 55; emphasis in the original.
14. Matthews, *"Just a Housewife,"* 111.
15. Margolis, *Mothers and Such: Views of American Women and Why They Changed*, 13.
16. Jeanne Boydston, *Home & Work: Housework, Wages, and the Ideology of Labor in the Early Republic* (New York: Oxford University Press, 1990), xi.
17. Dolores Hayden, *The Grand Domestic Revolution* (Cambridge, MA: MIT Press, 1982), 299.
18. Lisa Hix, "Out of the Shadow of Aunt Jemima: The Real Black Chefs Who Taught Americans to Cook," *Collectors Weekly*, January 22, 2016, https://www.collectorsweekly.com/articles/out-of-the-shadow-of-aunt-jemima-the-real-black-chefs-who-taught-americans-to-cook/.
19. Janet A. Flammang, *The Taste for Civilization: Food, Politics, and Civil Society* (Urbana: University of Illinois Press, 2009), 43.
20. U.S. Census Bureau Statistics 2013, money.cnn.com/2013/01/31/news/economy/secretary-women-jobs/index.html?iid-HP_LN.
21. Flammang, *Taste for Civilization*, 133, 134.
22. Bruce German, interview by author, May 27, 2016, Davis, California.
23. Matthews, *"Just a Housewife,"* 151; Flammang, *Taste for Civilization*, 154.
24. Boydston, *Home & Work*, xix.
25. John Kenneth Galbraith, *Economics and the Public Purpose* (Boston: Houghton Mifflin, 1973), 233.

CHAPTER TEN: WE CAN'T EAT OUR WAY OUT OF THIS

1. Sabrina Tavernise, "FDA Makes It Official: BPA Can't Be Used in Baby Bottles and Cups," *The New York Times*, July 17, 2012, http://www.nytimes.com/2012/07/18/science/fda-bans-bpa-from-baby-bottles-and-sippy-cups.html.
2. Irene Tung et al., *The Growing Movement for $15* (New York: National Employment Law Project, November 2015), http://www.nelp.org/content/uploads/Growing-Movement-for-15-Dollars.pdf.
3. "Women Deserve Equal Pay," National Organization of Women, https://now.org/resource/women-deserve-equal-pay-factsheet/ (accessed January 10, 2018).
4. Michelle Ye Hee Lee, "Do 'Welfare' Recipients Get $35,000 in Benefits a Year?" *The Washington Post*, December 5, 2014, https://www.washingtonpost.com/news/fact-checker/wp/2014/12/05/grothman-single-parents-welfare/.
5. "Transforming Our Food System, a Conversation with Mark Bittman, Ricardo Salvador, and Emily Broad Leib," moderated by Kat Taylor, Harvard Law School, November 30, 2016, YouTube video recording, https://www.youtube.com/watch?v=eQlK11BmE8k.
6. Maria Shriver, *The Shriver Report: A Woman's Nation Pushes Back from the Brink*, ed. Olivia Morgan and Karen Skelton (New York: Rosetta, 2013), time.com/2026/11-surprising-facts-about-women-and-poverty-from-the-shriver-report/.
7. See: Ben Schiller, "Everything You Need to Know About Universal Basic Income in 10 Minutes," *Fast Company*, December 12, 2017, https://www.fastcompany.com/40505620/everything-you-need-to-know-about-universal-basic-income-in-10-minutes; Martin Sandbu, "Universal Basic Income: Renaissance for a 500-Year-Old Idea," *Financial Times*, December 8, 2017, https://www.ft.com/content/3b7938e6-c569-11e7-b30e-a7c1c7c13aab; Annie Lowrey, "Switzerland's Proposal to Pay People for Being Alive," *The New York Times Magazine*, November 12, 2013, http://www.nytimes.com/2013/11/17/magazine/switzerlands-proposal-to-pay-people-for-being-alive.html; Peter S. Goodman, "Free Cash in Finland. Must Be Jobless," *The New York Times*, December 17, 2016, https://www.nytimes.com/2016/12/17/business/economy/universal-basic-income-finland.html.

8. Kristin Wartman, "Pay People to Cook at Home," *The New York Times*, op-ed, May 10, 2013, http://www.nytimes.com/2013/05/11/opinion/pay-people-to-cook-at-home.html.
9. Charlotte Alter, "11 Surprising Facts About Women and Poverty from the Shriver Report," *Time*, January 13, 2014, http://time.com/2026/11-surprising-facts-about-women-and-poverty-from-the-shriver-report/.
10. Power of Women Collective, foreword for Mariarosa Dalla Costa and Selma James, *The Power of Women and the Subversion of the Community* in Selma James, *Sex, Race, and Class—The Perspective of Winning: A Selection of Writings, 1952–2011*, 3rd ed. (Oakland, CA: PM Press, 2012), 45; James, introduction to *Power of Women*, 51; Della Costa and James, *Power of Women*, 59; emphases in the original.
11. Jennifer J. Deal, "Welcome to the 72-Hour Work Week," *The Harvard Business Review*, September 12, 2013, https://hbr.org/2013/09/welcome-to-the-72-hour-work-we.
12. John Holmberg, et al., *Low-Carbon Transitions and the Good Life, Report 6495* (Stockholm: Swedish Environmental Protection Agency, 2012), https://www.naturvardsverket.se/Documents/publikationer6400/978-91-620-6495-2.pdf.
13. Anders Hayden, "Europe's Work-Time Alternatives," in *Take Back Your Time: Fighting Overwork and Time Poverty in America*, ed. John de Graaf (Oakland, CA: Berrett-Koehler, 2003): 202–10.
14. Alexander Starritt, "Women and Men Use Cities Very Differently," Co.Design, December 7, 2016, https://www.fastcodesign.com/3066264/women-and-men-use-cities-very-differently.
15. Svati Narula, "Why Did Only 2% of New Fathers in Japan Take Paternity Leave Last Year?," *World Economic Forum*, January 11, 2016, https://www.weforum.org/agenda/2016/01/why-did-only-2-of-new-fathers-in-japan-take-paternity-leave-last-year/.
16. The OECD Better Life Index, "Obesity Update 2017," www.oecd.org/health/obesity-update.htm; U.S. Centers for Disease Control and Prevention, "Adult Obesity Facts," August 29, 2017, https://www.cdc.gov/obesity/data/adult.html. The U.S. obesity rate for Latinos is 42.5 percent and 47.8 percent for African Americans, according to the CDC.
17. Lena Graber and John Miller, "Wages for Housework: The Movement and the Numbers," *Dollars and Sense* (September/October 2002): 45–46.

18. H.R. 3486—RISE Out of Poverty Act, Session of 2015–6 (114th U.S. Congress), https://www.congress.gov/bill/114th-congress/house-bill/3486.
19. H.R. 4379—Women's Option to Raise Kids Act, Session of 2011–2 (112th U.S. Congress), https://www.congress.gov/bill/112th-congress/house-bill/4379.
20. Alex Emmons, "The Senate's Military Spending Increase Alone Is Enough to Make Public College Free," *The Intercept*, September 18, 2017.
21. Jeff Daniels, "Latest Budget Showdown Worries Pentagon, Risks Delaying Trump's Military Buildup," *CNBC.com*, January 17, 2018.
22. Trust for America's Health and the Robert Wood Johnson Foundation, "The Healthcare Costs of Obesity," The State of Obesity, no date, https://stateofobesity.org/healthcare-costs-obesity/ (accessed September 10, 2017).
23. American Heart Association, "Heart Disease and Stroke Statistics—at-a-Glance," December 17, 2014, https://www.heart.org/idc/groups/ahamah-public/@wcm/@sop/@smd/documents/downloadable/ucm_470704.pdf; American Cancer Society, *Cancer Facts & Figures 2014* (Atlanta: American Cancer Society, 2014), https://www.cancer.org/content/dam/cancer-org/research/cancer-facts-and-statistics/annual-cancer-facts-and-figures/2014/cancer-facts-and-figures-2014.pdf; American Diabetes Association, "American Diabetes Association Releases New Research Estimating Annual Cost of Diabetes at $245 billion," press release, March 6, 2013, http://www.diabetes.org/newsroom/press-releases/2013/annual-costs-of-diabetes-2013.html.
24. Melissa Bartick and Arnold Reinhold, "The Burden of Suboptimal Breastfeeding in the United States: A Pediatric Cost Analysis," *Pediatrics*, April 2010.
25. Joel Salatin, *Folks, This Ain't Normal* (New York: Center Street, 2011), 77.
26. Will Allen, *The Good Food Revolution* (New York: Gotham, 2012), 185.
27. UN News Service, "UN Report: One-Third of World's Food Wasted Annually, at Great Economic, Environmental Cost," *UN News Centre*, September 11, 2013, http://www.un.org/apps/news/story.asp?NewsID=45816#.WgUNX5OGNE4.
28. "UN: Global Food Waste Emissions Greater Than US Transport Sector," *Climate Change News*, November 9, 2013, http://www.climatechange

news.com/2013/09/11/un-global-food-waste-emissions-greater-than-us-transport-sector/.

29. "What We Grow," Growing Power, http://www.growingpower.org/education/what-we-grow/ (accessed June 15, 2017). In late 2017, the board of directors dissolved Growing Power, but Allen says he has no plans of retiring.
30. Silvia Federici, *Revolution at Point Zero: Housework, Reproduction, and Feminist Struggle* (Oakland, CA: PM Press, 2012), 15–16.
31. Kristin Wartman, "Allen Zimmerman: Produce Zen Master," *Linewaiters' Gazette*, August 7, 2014, https://www.foodcoop.com/files_lwg/lwg_2014_08_07_vII_n16.pdf. The gazette is the Park Slope Coop's newsletter.
32. Joe Holtz, email to author, December 14, 2016.
33. Nina Lopez, introduction to *Sex, Race, Class* (Oakland, CA: PM Press, 2012), 8.
34. Selma James, interview by author, March 20, 2014, New York, New York.
35. Ibid.
36. Adrian Parr, "Our Crime Against the Planet, and Ourselves," *The New York Times*, May 18, 2016, https://www.nytimes.com/2016/05/18/opinion/our-crime-against-the-planet-and-ourselves.html.

CONCLUSION: PROTECTING OUR BODIES AND OUR FUTURE

1. U.S. Centers for Disease Control and Prevention, "Mortality in the United States, 2015," December 2016, https://www.cdc.gov/nchs/data/databriefs/db267.pdf (accessed August 8, 2017).
2. Jay S. Olshansky et al., "A Potential Decline in Life Expectancy in the United States in the 21st Century," Special Report to *The New England Journal of Medicine*, March 17, 2005, http://www.nejm.org/doi/full/10.1056/NEJMsr043743#t=article.
3. My thinking here was influenced by Jonathan Latham, "Why the Food Movement Is Unstoppable," *Independent Science News*, September 20, 2016, https://www.independentsciencenews.org/health/why-the-food-movement-is-unstoppable/.

A NEW FOOD MOVEMENT MANIFESTO

1. Public Health Law Center, "Master Settlement Agreement," 2017, http://www.publichealthlawcenter.org/topics/tobacco-control/tobacco-control-litigation/master-settlement-agreement.
2. "Nixon Signs Bill Banning Radio-TV Cigarette Ads," *The New York Times*, April 2, 1970, http://www.nytimes.com/1970/04/02/archives/nixon-signs-bill-banning-radiotv-cigarette-ads.html.
3. Ibid.
4. Jodi Helmer, "Will This New Bill Level the Playing Field for Urban Farms?" *Civil Eats*, October 13, 2016, https://civileats.com/2016/10/13/will-this-new-bill-level-the-playing-field-for-urban-farms/.
5. Michelle Chen, "Could a Universal Basic Income Work in the US?" *The Nation*, August 15, 2017, https://www.thenation.com/article/could-a-universal-basic-income-work-in-the-us/.

INDEX